Orbital Mechanics
Second Edition

Orbital Mechanics
Second Edition

Edited by
Vladimir A. Chobotov

EDUCATION SERIES
J. S. Przemieniecki
Series Editor-in-Chief
Air Force Institute of Technology
Wright-Patterson Air Force Base, Ohio

Published by
American Institute of Aeronautics and Astronautics, Inc.
1801 Alexander Bell Drive, Reston, Virginia 20191–4344

American Institute of Aeronautics and Astronautics, Inc., Reston, Virginia

Library of Congress Cataloging-in-Publication Data

Orbital mechanics / edited by Vladimir A. Chobotov.—2nd ed.
    p.   cm.—(AIAA education series)
   Includes bibliographical references and index.
   1. Orbital mechanics.   2. Artificial satellites—Orbits.
3. Navigation (Astronautics).   I. Chobotov, Vladimir A.   II. Series
TL1050.O73   1996   629.4'113—dc20   96-42189
ISBN 1-56347-179-5

Third Printing.

Texts Published in the AIAA Education Series

Orbital Mechanics, Second Edition
    V. A. Chobotov, editor, 1996
Thermal Structures for Aerospace Applications
    Earl A. Thornton, 1996
Structural Loads Analysis for Commercial Transport Aircraft:
    Theory and Practice
    Ted L. Lomax, 1996
Helicopter Flight Dynamics: The Theory and Application of Flying Qualities and
Simulation Modeling
    Gareth Padfield, 1996
Flying Qualities and Flight Testing of the Airplane
    Darrol Stinton, 1996
Flight Performance of Aircraft
    S. K. Ojha, 1995
Operations Research Analysis in Quality Test and Evaluation
    Donald L. Giadrosich, 1995
Radar and Laser Cross Section Engineering
    David C. Jenn, 1995
Introduction to the Control of Dynamic Systems
    Frederick O. Smetana, 1994
Tailless Aircraft in Theory and Practice
    Karl Nickel and Michael Wohlfahrt, 1994
Mathematical Methods in Defense Analyses
    Second Edition
    J. S. Przemieniecki, 1994
Hypersonic Aerothermodynamics
    John J. Bertin, 1994
Hypersonic Airbreathing Propulsion
    William H. Heiser and David T. Pratt, 1994
Practical Intake Aerodynamic Design
    E. L. Goldsmith and J. Seddon, Editors, 1993
Acquisition of Defense Systems
    J. S. Przemieniecki, Editor, 1993
Dynamics of Atmospheric Re-Entry
    Frank J. Regan and Satya M. Anandakrishnan, 1993
Introduction to Dynamics and Control of Flexible Structures
    John L. Junkins and Youdan Kim, 1993
Spacecraft Mission Design
    Charles D. Brown, 1992
Rotary Wing Structural Dynamics and Aeroelasticity
    Richard L. Bielawa, 1992
Aircraft Design: A Conceptual Approach
    Second Edition
    Daniel P. Raymer, 1992
Optimization of Observation and Control Processes
    Veniamin V. Malyshev, Mihkail N. Krasilshikov, and Valeri I. Karlov, 1992
Nonlinear Analysis of Shell Structures
    Anthony N. Palazotto and Scott T. Dennis, 1992
Orbital Mechanics
    Vladimir A. Chobotov, 1991
Critical Technologies for National Defense
    Air Force Institute of Technology, 1991
Defense Analyses Software
    J. S. Przemieniecki, 1991

(Continued on next page.)

Texts Published in the AIAA Education Series (continued)

Inlets for Supersonic Missiles
    John J. Mahoney, 1991
Space Vehicle Design
    Michael D. Griffin and James R. French, 1991
Introduction to Mathematical Methods in Defense Analyses
    J. S. Przemieniecki, 1990
Basic Helicopter Aerodynamics
    J. Seddon, 1990
Aircraft Propulsion Systems Technology and Design
    Gordon C. Oates, Editor, 1989
Boundary Layers
    A. D. Young, 1989
Aircraft Design: A Conceptual Approach
    Daniel P. Raymer, 1989
Gust Loads on Aircraft: Concepts and Applications
    Frederic M. Hoblit, 1988
Aircraft Landing Gear Design: Principles and Practices
    Norman S. Currey, 1988
Mechanical Reliability: Theory, Models and Applications
    B. S. Dhillon, 1988
Re-Entry Aerodynamics
    Wilbur L. Hankey, 1988
Aerothermodynamics of Gas Turbine and Rocket Propulsion,
    Revised and Enlarged
    Gordon C. Oates, 1988
Advanced Classical Thermodynamics
    George Emanuel, 1988
Radar Electronic Warfare
    August Golden Jr., 1988
An Introduction to the Mathematics and Methods of Astrodynamics
    Richard H. Battin, 1987
Aircraft Engine Design
    Jack D. Mattingly, William H. Heiser, and Daniel H. Daley, 1987
Gasdynamics: Theory and Applications
    George Emanuel, 1986
Composite Materials for Aircraft Structures
    Brian C. Hoskins and Alan A. Baker, Editors, 1986
Intake Aerodynamics
    J. Seddon and E. L. Goldsmith, 1985
Fundamentals of Aircraft Combat Survivability Analysis and Design
    Robert E. Ball, 1985
Aerothermodynamics of Aircraft Engine Components
    Gordon C. Oates, Editor, 1985
Aerothermodynamics of Gas Turbine and Rocket Propulsion
    Gordon C. Oates, 1984
Re-Entry Vehicle Dynamics
    Frank J. Regan, 1984

Published by
American Institute of Aeronautics and Astronautics, Inc., Washington, DC

Foreword

The second edition of *Orbital Mechanics* by V. A. Chobotov et al. complements five other space related texts published in the Education Series of the American Institute of Aeronautics and Astronautics (AIAA); *Re-Entry Vehicle Dynamics* by F. J. Regan, *Introduction to the Mathematics and Methods of Astrodynamics* by R. H. Battin, *Space Vehicle Design* by M. D. Griffin and J. R. French, *Spacecraft Mission Design*, and *Spacecraft Propulsion*, both by C. D. Brown. The revised text in *Orbital Mechanics* is specifically designed as a teaching textbook with a significant amount of reference material and problems for the practicing aerospace engineer, scientist, or mission planner. It covers both the theory and applications of Earth orbits and interplanetary trajectories, including orbital maneuvers, space rendezvous, orbit perturbations, and collision hazards associated with space debris. The second edition includes two new chapters "Optimal Low-Thrust Orbit Transfers" and "Orbital Coverage," as well as expanded material on short- and long-term modeling of space debris. It also includes computer software to solve selected problems discussed in this textbook – a useful trend particularly encouraged in the AIAA Eduction Series.

The AIAA Education Series embraces a broad spectrum of theory and application of different disciplines in aerospace design practice. More recently the Series has been expanded to include defense science, engineering, and technology. The Series has been in existence for over ten years and its fundamental philosophy to develop texts that serve as both teaching texts for students and reference materials for practicing engineers and scientists has remained unchanged.

J. S. Przemieniecki
Editor-in-Chief
AIAA Education Series

Preface

The favorable reception of the First Edition of this volume appears to have sustained the author's belief in the need for a book written for the student of orbital mechanics with the emphasis on the applied, or engineering, aspects of the subject matter. After a brief review of the fundamentals, the book addresses the topics of orbital maneuvers, relative motion, orbital perturbations, orbital systems, lunar and interplanetary trajectories, and the contemporary topic of space debris. The keynote of the book is the practical utility of the material presented for use by the student or the practicing engineer.

In preparing the Second Edition, the authors have added two new chapters on orbital coverage and on optimal low-thrust orbit transfers. The first topic is of interest in relation to many classes of satellites whose primary function involves coverage of the Earth's surface. The second topic, which has received a great deal of attention in the technical literature in the past several decades, deals with the theory, and examples, of optimal orbit transfers using low thrust. Many examples illustrate the underlying principles and provide information useful to the practicing engineer or scientist.

A considerable portion of the material contained in the First Edition has been updated and revised, such as Chapter 13, for example, dealing with space debris. Some new material on relative motion in orbit (Chapter 7) and orbital systems (Chapter 11) has been provided.

Finally, the IBM PC or compatible user-friendly software is appended, which can be used to solve selected problems posed in the text. Included are programs based on the Lambert solution and several integration approaches, such as the Adam's methods with error control, the Gauss–Jackson with $J_2 \cdots J_6$, the Runge Kutta fourth order with J_2, and the moon-Earth trajectory calculations based on the patched conic solution. It is hoped that this computational capability will facilitate the solution of selected problems for the student and aid in the evaluation of more complex engineering problems for the professional in the field.

V. A. Chobotov
July 1996

1
Basic Concepts

1.1 A Historical Perspective

One of man's earliest reasons for attempting to understand the motions of the sun, moon, and planets was his belief that they controlled his destiny. Other reasons were his need to measure time and later to use the celestial objects for navigation. Thus, the names assigned to the days of the week are closely related to the names of the celestial bodies: Saturn, Jupiter, Mars, sun, Venus, Mercury, and moon; by taking the first name and skipping two and repeating in this way, we have a partial derivation of the names of the week in French (*mardi, mercredi, jeudi, vendredi* for (Tuesday–Friday) or English (Saturday, Sunday, Monday).

The earliest evidence of man's interest in the universe dates back to 1650 B.C. in Babylon and Egypt (e.g., the Ahmes Papyrus). This evidence shows an elaborate system of numeration in which positional or place-value notation was used. For example, clay tablets with cuneiform writing show that the following sexagesimal (base 60) system notation was used:

$$P = 1, \qquad PP = 2, \qquad PPP = 3$$

$$\prec = 10 \qquad \prec\prec P = 2 \times 10 + 1 = 21$$

But

$$P\prec\prec = 1, 20$$
$$= 60 + 20 = 80$$

or

$$P\prec = 60 + 10 = 70, \text{ etc.}$$

Thus, in this notation, 3; 3,45 would mean $3 + 3/60 + 45/3600 = 3.0625$, etc. This system of numeration remained in use through the time of Copernicus, who would write 2, 9: 17, 22, 36 to mean 129th day ($2 \times 60 + 9$), 17th h, 22nd min, and 36th s. The sexagesimal system is, of course, still the basis for modern timekeeping (i.e., 60 s/min, etc.).

In 300 B.C., Aristarchus developed a theory in which the sun and stars were fixed and the Earth revolved in a circular orbit about the sun. Unfortunately, the leading philosophers of the time treated Aristarchus' theory with contempt. The most popular theory was that the Earth was a fixed center of the universe and that the planets moved around the Earth. In 130 B.C., Hipparchus introduced the epicyclical motion of the planets, which was further developed by Ptolemy in A.D. 150 as the principal theory for predicting the motions of the planets. Although there were no physical principles on which to base the motions, some of the results obtained by this theory (e.g., the rise and set of the planets) were very accurate. This picture remained virtually unchanged through the Middle Ages.

1

It is interesting to note that the early Greeks speculated and theorized about the size, shape, and composition of the Earth. Pythagoras and Aristotle supported the spherical figure for the Earth. Anaximenes, however, believed strongly that the Earth was rectangular.

Problems of geometry (e.g., finding areas of surfaces) required the knowledge of π. The Babylonians considered the area of a circle to be $(8/9D)^2$, where D was the diameter. This corresponds to $\pi \approx 3.16049$, which was not as accurate as the figure arrived at by Archimedes, who was the first to have a method for calculating π to any desired degree of accuracy. It was based on the fact that the perimeter of a regular polygon of n sides inscribed in a circle is smaller than the circumference of the circle, whereas the perimeter of a similar polygon circumscribed about the circle is greater than its circumference. By making n sufficiently large, the two perimeters will closely approach the circumference of the circle arbitrarily. Archimedes started with a hexagon and, progressively doubling the number of sides up to 96, obtained

$$3\frac{10}{71} < \pi < 3\frac{1}{7}$$

which was far better than that of the Babylonians. The Babylonians also regarded the perimeter of a hexagon as being equal to six times the radius of the circumscribed circle, which appears to be the reason they chose to divide the circle into 360 deg, a rule that we still live by.

Early in the sixteenth century, Copernicus (1467–1543) put forward a scheme that put the sun at the center of the universe. The planets were seen as moving in epicycles around the sun, with the moon moving around the Earth. Copernicus also hypothesized that the stars lay on a sphere of very large radius. His theory was not very well received in his day.

In 1601, Johann Kepler (1571–1630) became the director of the Prague Observatory on the death of Tycho Brahe (1546–1601), who had observed for 13 years the relative motion of the planet Mars. By 1609, Kepler had formulated his first two laws and, in 1619, published the third law, which he dedicated to James I of England. Kepler stated that

1) Every planet moves in an orbit that is an ellipse, with the sun at one focus of the ellipse.

2) The radius vector drawn from the sun to any planet sweeps out equal areas in equal times.

3) The squares of the periods of revolution of the planets are proportional to the cubes of the semimajor axes of their orbits.

As the result of Galileo's (1564–1642) observations of the four moons of Jupiter, the Copernican heliocentric theory was accepted, and Newton's theory of gravitation produced a theoretical principle that explained the motions of the planets and laid the foundation for modern space flight.

Isaac Newton (1642–1727)

Although unexceptional as a child, Newton began to produce immediately following his studies at Trinity College, Cambridge, in 1665. In that year, the plague was ravishing Cambridge, and Newton retired to the countryside to work in safety. He had already devised the binomial theorem and was beginning to grasp the calculus when the apple incident occurred (the tale sounds like an incredible legend, but it's true—it comes from Newton's own records).

Newton formulated his famous inverse-square law to describe the behavior of celestial bodies. Furthermore, he intuitively theorized that the full force of the Earth's gravity could be considered as emanating from a point source at its center. But when he checked his theory with calculations using an imprecise figure for the Earth's radius, he found he was off by enough to make him doubt his point-source assumption, and he set aside the problem indefinitely.

It was also during this two-year absence from Cambridge that he performed experiments with prisms, proving that white light is actually a combination of all the colors of the rainbow. The prism experiments made Newton famous, and he returned to Cambridge in 1667 as a Fellow of Trinity College. In 1669, when Newton was just 27, the Cambridge mathematics lecturer Isaac Barrows resigned in his student's favor, and Newton was appointed to the Lucasian Chair of Mathematics.

Three years later, Newton was elected to the Royal Society, to which he promptly reported all his findings in light. Although his papers quickly brought him international recognition, they also embroiled him in troublesome controversy, chiefly with Robert Hooke, who had performed some similar but far less brilliant optical experiments.

Hooke was not Newton's only adversary. Leibnitz developed the calculus simultaneously with, but quite independently of, Newton. Although the two men were friends, mathematicians and scientists in Germany and England soon fell into a chauvinistic debate over the true origin. Secretly, Newton, who was neurotically sensitive to any criticism, urged his supporters on.

His distinguished contributions in optics and mathematics notwithstanding, Newton's greatest discovery by far was the universal law of gravitation. In 1684, Hooke boasted to Wren and Halley that he had worked out the laws governing the motions of the heavenly bodies. Wren was intrigued with the problem but unimpressed with Hooke's explanation, and so he immediately put up a prize for the correct solution. Halley, a friend of Newton, took the problem to his colleague and asked how the planets would move if there was a force between all bodies decreasing as the square of the distance between them. Without a pause, Newton replied, "In ellipses." When Halley asked him how he knew, Newton said, "Why, I have calculated it." Then Newton told Halley about his earlier work during the plague years. Halley urged Newton to try his calculations once again. This time, Newton had a much more accurate figure for the radius of the Earth, and he knew from his calculations that the point-source assumption was correct.

Newton began to record his discoveries, and the result was *Philosophiae Naturalis Principia Mathematica*. One of the greatest scientific works ever published, *Principia Mathematica* climaxed the scientific revolution begun by Copernicus. The three laws of motion were established and, from them, the law of gravitation.

Afterward, honors came to Newton: elected to Parliament in 1689; appointed to the prestigious position of Warden of the Mint in 1696; elected president of the Royal Society in 1703; knighted in 1705.

But decades of bitter controversy and his own neurotic temperament took their toll in his later life. At one point he collapsed from a nervous breakdown and was forced to retire for two years. He had always been an ardent believer in alchemy and, in his later years, he wasted much of his time chasing recipies for gold. He became a mystic, too, producing a vast amount of writing on the more abstruse passages of the Bible.

Yet, in the end, he was a modest man, who said of his profound contributions, "I seem to have been only like a boy playing on the seashore, and diverting myself in now and then finding a smoother pebble or a prettier shell, while the great ocean of truth lay all undiscovered before me."

Sir Isaac Newton died in 1727 and was buried in Westminster Abbey along with the greatest of England's heroes.

Newton's Laws

The following three laws of motion given by Newton are considered the axioms of mechanics:

1) Every particle persists in a state of rest or of uniform motion in a straight line (i.e., with constant velocity) unless acted on by a force.

2) If F is the (external) force acting on a particle of mass m which, as a consequence, is moving with velocity v, then

$$F = \frac{d}{dt}(mv) = \frac{dp}{dt} \tag{1.1}$$

where $p = mv$ is called the momentum. If m is independent of time t, this becomes

$$F = m\frac{dv}{dt} = ma$$

where a is the acceleration of the particle.

3) If particle 1 acts on particle 2 with a force F_{12} in a direction along the line joining the particles while particle 2 acts on particle 1 with a force F_{21}, then $F_{21} = -F_{12}$. In other words, to every action there is an equal and opposite reaction.

It is assumed that m is a positive quantity and that it is large compared to the atomic particles. Furthermore, velocity v is regarded as small compared to the velocity of light (in nonrelativistic mechanics only).

Newton's laws define a force F in terms of the mass, velocity, and time (derivative with respect to time). Newton's definition of mass can be presented as follows:

> If two bodies A and B, which differ in material, size, and shape, give the same results when used in any experiment performed first on A and then on B, they are mechanically equivalent. The mass is a number assigned to a body A based on a comparison of mechanical equivalency with another body B. A distinction can be made between a gravitational mass of a body obtained by weighing and an inertial mass obtained from experiments (such as the measurement of F/acceleration) not affected by the gravitational fields. The equivalence of the inertial and gravitational masses was a subject of investigation by scientists from Newton (who found them equivalent to one part in 10^3) to Baron von Eotvos and R. H. Dicke. Eotvos performed his basic experiment in 1890 with a torsional balance, in which he determined the equivalency of the two masses to within the accuracy of his experiment (one part in 10^8). Dicke used a more refined apparatus,

employing three bodies, and found the equivalency to be one part in 10^{10}. The equivalency of the inertial and gravitational masses is, of course, the reason why all freely falling bodies have the same acceleration, as is required by the principle of equivalence in the general theory of relativity.

It must be emphasized that Newton's laws are postulated under the assumption that all measurements or observations are taken with respect to an "inertial" coordinate system or frame of reference that is fixed in space or is moving with a constant velocity (but not rotating). This is the so-called assumption that space or motion is absolute. It is quite clear, however, that a particle can be at rest or in uniform motion in a straight line with respect to one frame of reference and be traveling in a curve and accelerating with respect to another frame of reference.

The Earth is not exactly an inertial system but, for many practical purposes, it can be considered as one so long as motion takes place with speeds that are not too large. For speeds comparable to the speed of light (300,000 km/s), Newton's laws of mechanics must be replaced by Einstein's laws of relativity of relativistic mechanics.

1.2 Velocity and Acceleration

Kinematics and dynamics are two branches of physics that are concerned with the motion of material bodies and the effects produced on them by forces. Kinematics is the study of the geometry of motion, whereas dynamics is concerned with the physical causes of motion. The axiomatic foundations of these two branches of physics (mechanics) are based on

1) Undefined terms or concepts, such as a line in the Euclidian geometry.
2) Postulates or assertions based on experimental observations, such as Kepler's laws.
3) Definitions (displacement, velocity, accelerations, etc.).
4) Theorems or proved assertions based on definitions and postulates.

The study of the kinematics and dynamics of motion usually requires a reference coordinate frame in which the position, velocity, and acceleration of a material mass point can be specified. A nonrotating and nonaccelerating reference frame is known as the inertial frame in which the laws of mechanics are valid and can be most conveniently expressed. If an orthogonal inertial frame is used to denote the vector position r of a point, then its velocity and accelerations are $v = dr/dt$ and $a = dv/dt$, respectively. If the position r is measured in a frame that has an angular velocity ω and the translational velocity v_0, then the absolute or inertial velocity of the point is

$$v_{\text{inertial}} = \dot{r}' + \omega \times r + v_0 \qquad (1.2)$$

where $\dot{r}' = $ translational velocity of the point relative to the rotating frame.

The acceleration of the point in inertial space is then

$$a = \frac{dv}{dt}$$

$$= \ddot{r}' + \omega \times \dot{r}' + \dot{\omega} \times r + \omega \times (\dot{r}' + \omega \times r) + \dot{v}_0$$

$$= \ddot{r}' + 2\omega \times \dot{r}' + \dot{\omega} \times r + \omega \times (\omega \times r) + a_0 \qquad (1.3)$$

where $\ddot{r}'$ = observed acceleration of the point relative to the rotating frame, and a_0 = acceleration acceleration of the rotating frame.

The second and fourth terms in Eq. (1.3) are the Coriolis and centripetal accelerations, respectively. The $\dot{\omega} \times r$ term is the tangential (Euler) acceleration. The Coriolis acceleration $a_c = 2\omega \times \dot{r}'$ term exists only when there is an observed velocity $\dot{r}'$ in a rotating frame.

Projectile on a Rotating Earth

The coordinate frames convenient for the use of Eq. (1.3) to compute the Coriolis acceleration terms for a particle ejected with a velocity $\dot{r}'$ at latitude λ on the Earth's surface are illustrated in Fig. 1.1.

The orthogonal frame E_1, E_2, E_3 is an Earth-centered inertial (ECI) frame in Fig. 1.1. The e_1, e_2, e_3 is a frame rotating with the Earth angular velocity ω, in which the particle p is given an easterly velocity $\dot{r}'$ at an elevation angle α. The acceleration of the particle p relative to the E_1, E_2, E_3 frame is

$$a = a_0 + \ddot{r}' + 2\omega \times \dot{r}' + \omega \times (\omega \times r)$$
$$= g = \text{gravitational acceleration} \tag{1.4}$$

Here,

$$a_0 = \omega \times (\omega \times R) = \text{centripetal acceleration of } 0' \text{ relative to } 0$$

$$\ddot{r}' = \text{apparent (relative) acceleration}$$

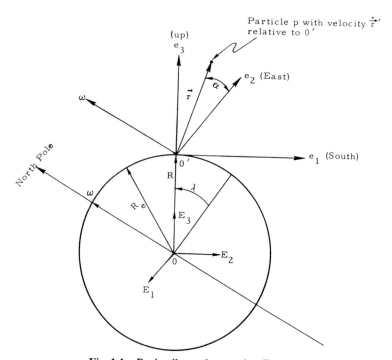

Fig. 1.1   Projectile on the rotating Earth.

If a_0 and $\omega \times (\omega \times r) \ll 2\omega \times \dot{r}'$, Eq. (1.4) can be solved approximately for the acceleration of p relative to $0'$ as

$$\ddot{r}' \approx g - 2\omega \times \dot{r}'$$

Rocket Equation

An important application of Newton's second law of motion is in the derivation of the rocket equation. For example, let a mass m with a velocity v collide inelastically with an element of mass Δm, which has a velocity v_1. Then, prior to impact, the initial system momentum (in scalar form) is

$$P_1 = mv + \Delta m v_1 \tag{1.5}$$

At a time Δt later, the two masses are joined, and their combined momentum is

$$P_2 = (m + \Delta m)(v + \Delta v) \tag{1.6}$$

Newton's laws state that the impulse is equal to the change in momentum or

$$\begin{aligned} \text{Impulse} &= F\Delta t \\ &= P_2 - P_1 \\ &= m\Delta v + \Delta m(v - v_1) + \Delta m \Delta v \end{aligned} \tag{1.7}$$

where, in the limit as $\Delta t \to 0$ and $\Delta m, \Delta v \to 0$, the force F can be expressed as

$$F = m\frac{dv}{dt} + \frac{dm}{dt}(v - v_1) \tag{1.8}$$

If $v_1 = 0$, then this result is equivalent to the case of a falling raindrop, for example, which absorbs moisture (stationary) from the air. If, on the other hand, $u = v - v_1$, then Eq. (1.8) can be written as

$$F = m\frac{dv}{dt} - T \tag{1.9}$$

where $T = -u(dm/dt) =$ momentum thrust of a rocket. Equation (1.9) is known as the rocket equation, where F is the external force acting on the rocket (e.g., gravity, aerodynamic, etc.), and u is the exhaust velocity.

A basic application of Eq. (1.9) is in the calculation of velocity impulses (ΔV) for rocket motion. A somewhat different form of this equation can be obtained by assuming that $F = 0$ and rewriting Eq. (1.9) as

$$\begin{aligned} dv &= \frac{T\,dt}{m} \\ &= \frac{-u\,dm}{m} \end{aligned}$$

which, after integration, becomes

$$
\begin{aligned}
\Delta v &= u \ln \frac{m_i}{m_f} \\
&= g_0 I_{sp} \ln \frac{m_i}{m_f}
\end{aligned}
\tag{1.10}
$$

where

m_i = initial mass of rocket

m_f = final mass of rocket

I_{sp} = propellant specific impulse (thrust/propellant weight flow rate)

g_0 = gravitational constant at sea level

Equation (1.10) can be used to compute the mass of propellant for a given value of ΔV. It is also known as the Tsiolkovsky formula.

The specific impulse I_{sp} is a parameter of propellant quality and generally varies from 60 to 3000 + s for cold gas or ion propellants, respectively. It is a measure of propellant thrust obtained per unit of propellant weight flow. For example, Eq. (1.10) can be solved for the final mass remaining after a burn of velocity impulse ΔV as

$$
m_f = m_i \exp(-\Delta V / g_0 I_{sp})
\tag{1.11}
$$

in terms of the initial mass m_i and the specific impulse I_{sp}.

If there are two or more impulses $\Delta V_i (i = 1, 2, \ldots, n)$ with a coast period between the impulses, the final mass remaining m_f is given by the equation

$$
m_f = m_i \exp(-\Delta V_T / g_0 I_{sp})
\tag{1.12}
$$

where

$\Delta V_T = \Delta V_1 + \Delta V_2 + \cdots + \Delta V_n$

$m_i \quad = m_{bo} + m_p + m_{pl} =$ initial rocket mass

$m_f \quad = m_{bo} + m_{pl} =$ final rocket mass

$m_{bo} \quad =$ burnout mass (structural mass)

$m_p \quad =$ propellant mass

$m_{pl} \quad =$ payload mass

The propellant mass ratio $\varepsilon = m_p / (m_p + m_{bo})$. The structural factor is $1 - \varepsilon$. The mass of the propellant burned is

$$
m_p = m_i [1 - \exp(-\Delta V_T / g_0 I_{sp})]
\tag{1.13}
$$

Example. Compute the propellant mass m_p required to deliver a payload of 1000 kg to a mission orbit requiring a velocity impulse $\Delta V = 4354$ m/s. Assume that the burnout mass of the rocket (structure) is 2000 kg and that the specific impulse of the propellant (LO_2/LH_2) is 460 s.

Using Eq. (1.11), the mass ratio

$$\frac{m_i}{m_f} = \frac{m_{pl} + m_{bo} + m_p}{m_{pl} + m_{bo}}$$

$$= \exp(\Delta V / g I_{sp})$$

Solving for $m_p = 4881$ kg.

Problems

1.1. A bar $AB = 3$ m is moving in a plane. At a given instant, the velocities of A and B are

$$V_A = 2 \text{ m/s, } 60 \text{ deg clockwise from line } A \text{ to } B$$

$$V_B = 30 \text{ deg clockwise from line } A \text{ to } B$$

Determine the angular velocity ω of the bar, stating whether it is clockwise or counterclockwise.

1.2. Determine the radial and normal acceleration components for a particle moving in a plane.

1.3. A single-stage rocket of mass $m_0 = m_{bo} + m_p$ carries a payload of mass m_{pl}. If the propellant mass ratio $\varepsilon = 0.8$, find the maximum payload that can be given a final speed of 1.3 u, where u is the exhaust velocity. Here $m_{bo} =$ burnout mass, and $m_p =$ propellant mass.

1.4. Determine the remaining propellant in the orbit-to-orbit shuttle after deployment of a 1000-kg payload to the circular synchronous equatorial orbit. Assume a start from, and a return to, a low-altitude circular parking orbit. Assume the following:
1) Ascent transfer-velocity impulse
   $\Delta V_1 = 2524$ m/s
2) Synchronous equatorial orbit injection-velocity impulse
   $\Delta V_2 = 1830$ m/s
3) Descent transfer-velocity impulse
   $\Delta V_3 = \Delta V_2$
4) Parking orbit injection velocity impulse
   $\Delta V_4 = \Delta V_1$
5) Propellant (LO_2/LH_2)
   $I_{sp} = 460$ s

Also, let

$m_{\text{bo}} = 3041$ kg = burnout mass
$m_p = 26500$ kg = propellant mass
$m_{\text{pl}} = 1000$ kg = payload mass

HINT: Given W_i, Find W_{p_2}.
Use

$$\frac{W_i}{W_1} = \exp[(\Delta V_1 + \Delta V_2)/g I_{\text{sp}}]$$

$$\frac{W_2}{W_f} = \exp[(\Delta V_3 + \Delta V_4)/g I_{\text{sp}}]$$

where

$$W_1 = W_{\text{bo}} + W_{p_1} + W_{\text{pl}}$$

$$W_2 = W_1 - W_{\text{pl}}$$

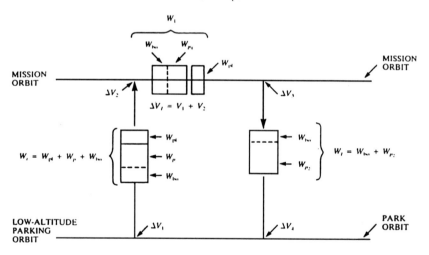

Selected Solutions

1.1. $\omega = 0.383$ rad/s counterclockwise

1.3. $m_{\text{pl}} = m_0/10$

1.4. $W_{p_2} = 1004$ kg

Celestial Relationships

2.1 Coordinate Systems

Several coordinate systems are used in the study of the motions of the Earth and other celestial bodies. The heliocentric-ecliptic coordinate system, illustrated in Fig. 2.1, has its origin at the center of the sun.[1] The $X_\varepsilon Y_\varepsilon$ fundamental plane coincides with the ecliptic plane, which is the plane of the Earth's revolution around the sun. The line of intersection of the ecliptic plane and the Earth's equatorial plane defines the direction of the X_ε axis. On the first day of spring, a line joining the center of the Earth and the center of the sun points in the direction of the positive X_ε axis. This is called the vernal equinox direction and is denoted by the symbol of a ram's head by astronomers because it points in the general direction of the constellation Aries. All Earth locations experience identical durations of daylight and darkness. The Earth's spin axis wobbles slightly and shifts in direction slowly over centuries. This effect is known as precession, and it causes the line of intersection of the Earth's equatorial plane and the ecliptic plane to shift slowly. As a result, the heliocentric-ecliptic system is not an inertial reference frame. Where extreme precision is required, it is necessary to specify the coordinates of an object based on the vernal equinox direction of a particular year or epoch.

ECI System

The geocentric-equatorial coordinate system, on the other hand, has its origin at the Earth's center. The fundamental plane is the equator, and the positive X axis points in the vernal equinox direction. The Z axis points in the direction of the North Pole. (At equinox all Earth locations experience identical durations of daylight and darkness.) This is shown in Fig. 2.2. It is important to keep in mind, when looking at Fig. 2.2, that the X, Y, Z system is not fixed to the Earth and turning with it; rather, the geocentric-equatorial frame is nonrotating with respect to the stars (except for precession of the equinoxes), and the Earth turns relative to it.

The two angles needed to define the location of an object along some direction from the origin of the celestial sphere are defined as follows:

α (right ascension) = the angle measured eastward in the plane of the equator from a fixed inertial axis in space (vernal equinox) to a plane normal to the equator (meridian), which contains the object; 0 deg $\leq \alpha \leq$ 360 deg.

δ (declination) = the angle between the object and equatorial plane measured (positive above the equator) in the meridional plane, which contains the object; -90 deg $\leq \delta \leq 90$ deg.

r (radial distance) = the distance between the origin of the coordinate system and the location of a point (object) within the coordinate system; $r \geq 0$.

To an observer fixed in an inertial frame (i.e., one that is at rest or moves with constant velocity relative to distant galaxies), the Earth revolves about the sun

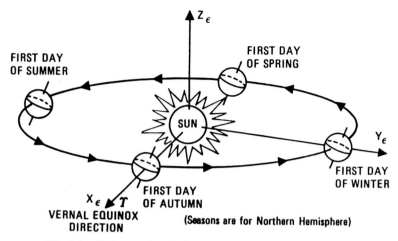

Fig. 2.1   Heliocentric ecliptic coordinate system (from Ref. 1).

in a nearly circular orbit in the ecliptic plane. The Earth rotates at an essentially constant rate about its polar axis. However, that axis of rotation is tilted away from the normal to the ecliptic. This tilt causes the equatorial and ecliptic planes to form a dihedral angle that is conventionally known as the obliquity of the ecliptic and is denoted by $\varepsilon(\approx23.5 \text{ deg})$.

On account of this angle between the two planes, the sun, as viewed by an observer on Earth, appears to shuttle northward and southward across the equatorial plane (called the celestial equator). This gives rise to the changing seasons during the course of the Earth's yearly revolution about the sun.

Of particular interest is the time each year when the sun moves northward through the equatorial plane; this event, called the vernal equinox, marks the beginning of spring. It is customary to associate a direction in inertial space with

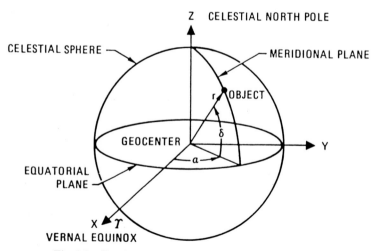

Fig. 2.2   Earth-centered inertial system (from Ref. 2).

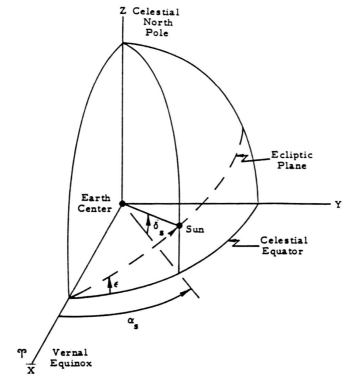

Fig. 2.3   Motion of the sun in the ECI reference frame (from Ref. 3).

this event by noting the position of the sun as seen from Earth against the field of infinitely distant background galaxies. Figure 2.3 illustrates the motion of the sun in the ECI frame frequently used in defining celestial relationships. Note that the X axis is directed toward the vernal equinox and that the Z axis lies along the Earth's polar axis of rotation; the Y axis completes the right-handed triad. The three arcs drawn in the three references planes bound an octant of the celestial sphere that is centered at the Earth and of arbitrary radius. The path of the sun's motion is shown by the dashed arc on the celestial sphere.

About three months after the vernal equinox, the sun reaches its northernmost position (and, at local noon, is directly overhead to observers on the Tropic of Cancer at about 23.5°N); this event is called the summer solstice, since the sun appears to stand still momentarily as it reverses its direction of motion from northward to southward. Then, about three months after the summer solstice, the sun crosses the celestial equator from north to south at the autumnal equinox. The sun reaches its southernmost position about three months later at the winter solstice, which marks the start of winter. At that time, the sun is directly overhead to observers, at local noon, on the Tropic of Capricorn at about 23.5°S.

Referring again to Fig. 2.3, it is customary to indicate the sun's position by a pair of angles similar to the familiar longitude and latitude system used for terrestrial measurements. The right ascension of the sun is the angle measured

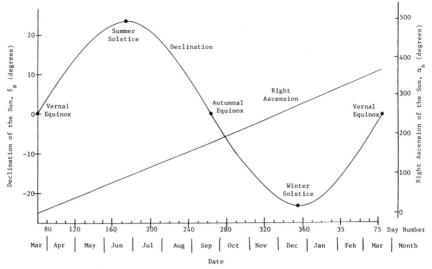

Fig. 2.4   Solar declination and right ascension vs date (from Ref. 3)

positively eastward along the celestial equator to the meridional plane (through the celestial polar axis), which passes through the sun; this angle, which is the counterpart of terrestrial longitude, is denoted as α_s. The second angle used to indicate the sun's position is the declination measured positively northward from the celestial equator; this angle, which is the counterpart of geocentric latitude, is denoted by δ_s. Figure 2.4 shows the variation with time of solar declination and right ascension.

Geographic Coordinate System

It is customary to locate an object relative to the Earth by two angular co-ordinates (latitude-longitude) and altitude above (or perhaps below) the adopted reference ellipsoid. The present discussion considers the Earth to be spherical. The origin of the latitude-longitude coordinate system is the geocenter (Fig. 2.5). The fundamental plane is the equator, and the principal axis in the fundamental plane points toward the Greenwich meridian.

The two angles required to define the location of a point along some ray from the geocenter are defined as follows:

ϕ (geocentric latitude) = the acute angle measured perpendicular to the equatorial plane between the equator and a ray connecting the geocenter with a point on the Earth's surface; -90 deg $\leq \phi \leq 90$ deg.

λ_E (east longitude) = the angle measured eastward from the prime meridian in the equatorial plane to the meridian containing the surface point; 0 deg $\leq \lambda_E \leq$ 360 deg.

The observer's coordinate frame is related to the previously discussed inertial frame through the angle Θ, the sidereal time; Θ is measured eastward in the equatorial plane from the vernal equinox to the observer's meridian and ranges from 0^h to 24^h (or, equivalently, from 0 to 360 deg). The distance from the geocenter to the object is called the range and is denoted by r.

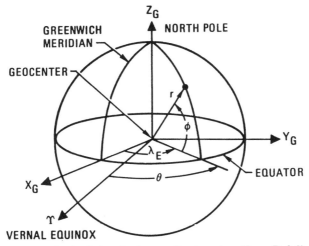

Fig. 2.5   Latitude-longitude coordinate system (from Ref. 2).

Azimuth-Elevation Coordinate System

An observer standing at a particular point on the surface of a rotating planet sees objects in a rotating coordinate system. In this system, the observer is at the origin of the system, and the fundamental plane is the local horizon (Fig. 2.6). Such a coordinate system is referred to as a topocentric system. Generally, the principal axis or direction is taken as pointing due south. Relative to an observer, the object is in a meridional plane that contains the object and passes through the zenith of the observer.

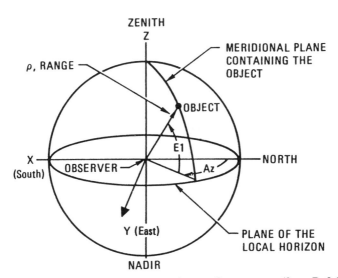

Fig. 2.6   Azimuth-elevation topocentric coordinate system (from Ref. 2).

Table 2.1   Constants and conversion factors

Earth Gravitational Constant:[a]

$$\mu = 5.53042885 \times 10^{-3} \text{ ER}^3 \text{ (equatorial Earth Radius)}^3/\text{min}^2$$
$$= 3.986005 \times 10^5 \text{ km}^3/\text{s}^2$$
$$= 1.40764438 \times 10^{16} \text{ ft}^3/\text{s}^2$$
$$= 6.27501680 \times 10^4 \text{ (n.mi.)}^3/\text{s}^2$$

Mean Equatorial Earth Radius:

$$\text{ER} = 6.37813700 \times 10^3 \text{ km}$$
$$= 3.44391847 \times 10^3 \text{ n.mi.}$$

Rotational Rate of the Earth:

$$\omega_E = 7.29211515 \times 10^{-5} \text{ rad/s}$$

Mean Obliquity of the Ecliptic (1 January 1976):

$$\varepsilon = 23°\ 26'\ 32.66''$$
$$= 23.442405°$$

Time Conversions:

1 mean solar day = 86,400 ephemeris s
1 mean sidereal day = 86164.09054 ephemeris s
1 tropical yr = 365.2421988 mean solar days
1 calendar yr = 365 mean solar days
1 mean solar s = 1.1574074×10^{-5} mean solar day
$\qquad\qquad\qquad = 1.1605763 \times 10^{-5}$ sidereal day
$\qquad\qquad\qquad = 1.002737$ sidereal s
Julian day = continuous count of the number of days since noon,
$\qquad\qquad$ 1 January 4713 B.C.

[a]IAU (1976) Astrodynamics Standards.

The two angles needed to define the location of an object along some ray from the origin are defined as follows:

Az (azimuth) = the angle, eastward from north to the object's meridian, as measured in the local horizontal plane, which is tangent to the sphere at the observer's position; 0 deg ≤ *Az* < 360 deg.

El (elevation) = angular elevation (measured positively upward in the meridional plane) of an object above the local horizontal plane, which is tangent to the sphere at the observer's position; −90 deg ≤ *El* ≤ 90 deg.

The distance from the observer to the object is called the range and is usually denoted by ρ. $\rho \geq 0$.

Coordinate Transformations

The satellite state vector at a given time is obtained by integrating the equations of motion that equate the acceleration of the vehicle to the sum of the various accelerations acting on the vehicle. The integration must be performed in an inertial (nonrotating) reference frame. However, the principal acceleration due to gravity and aerodynamic drag is expressed mostly in the rotating (body-fixed) systems. Transformations from body-fixed coordinates to inertial and back are therefore required. These transformations involve the motion of the equinox, which is due to the combined motions of the Earth's equatorial plane and the ecliptic plane, the equinox being defined as the intersection of these planes. The motion of the equatorial plane is due to the gravitational attraction of the sun and moon on the Earth's equatorial bulge. It consists of the lunisolar precession and nutation. The former is the smooth, long-period westward motion of the equator's mean pole around the ecliptic pole, with an amplitude of about 23.5 deg and a period of about 26,000 yr. The latter (nutation) is a relatively short-period motion that carries the actual (or true) pole around the mean pole in a somewhat irregular curve, with an amplitude of approximately 9 s of arc and a period of about 18.6 yr. The motion of the ecliptic (i.e., the mean plane of the Earth's orbit) is due to the planet's gravitational attraction on the Earth and consists of a slow rotation of the ecliptic. This motion is known as planetary precession and consists of an eastward movement of the equinox of about 12 s of arc a century and a decrease of the obliquity of the ecliptic, the angle between the ecliptic and the Earth's equator, of approximately 47 s of arc a century. The "true" equator and equinox are obtained by correcting the mean equator and equinox for nutation.

2.2 Time Systems

Astronomers also specify the location of stars by their right ascensions and declinations, which are essentially invariant. (Very gradual changes in star locations occur as a result of the precession of the equinoxes, which is caused by coning of the Earth's spin axis in inertial space.) The angular coordinates of the sun, however, vary considerably in the course of each year, as shown in Fig. 2.4 or by the tabulations of sun position given in a solar ephemeris. As indicated earlier, the sun moves at an irregular rate (as a result of noncircularity of the Earth's orbit about the sun) along the dashed arc in Fig. 2.3. The time between two successive passages through the vernal equinox is called a tropical year (about 365.25 days). Because of the precession of the equinoxes, the tropical year is about 20 min shorter than the orbital period of the Earth relative to the fixed-star or sidereal year.[4]

On account of the irregular motion of the true sun along the ecliptic, astronomers have introduced a fictitious mean sun, which moves along the celestial equator at a uniform rate and with exactly the same period as the true sun. It is relative to this fictitious mean sun that time is reckoned. This can best be explained by the geometry illustrated in Fig. 2.7, in which the celestial sphere is viewed by an observer at the celestial North Pole looking in the direction of the negative Z axis of the Earth-centered inertial (ECI) frame.

The rotational orientation of the Earth is conventionally determined by specification of the Greenwich hour angle of the vernal equinox (GHAVE), which is measured positively westward along the celestial equator from the Greenwich

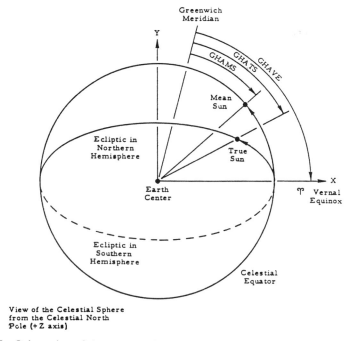

Fig. 2.7   Orientation of the mean and true sun and the Greenwich meridian in the ECI frame (from Ref. 3).

meridian to the vernal equinox. (Greenwich hour angle is measured either in degrees or in hours, the conversion factor being exactly 15 deg/h). Similarly, the instantaneous locations of the mean and the true suns are determined by Greenwich hour angle of the mean sun (GHAMS) and Greenwich hour angle of the true sun (GHATS), respectively. Incidentally, the difference (GHATS − GHAMS), also known as the equation of time, can be appreciated as the discrepancy between time reckoned by a sundial (based on the true sun position) and conventionally determined time (based on the position of the mean sun). Greenwich mean time (GMT) is linearly related to the GHAMS as follows[3]:

$$\text{GMT(h)} = [12 + \text{GHAMS(deg)}/15]\text{mod } 24 \qquad (2.1)$$

Noon GMT occurs when the mean sun is at upper transit of the Greenwich meridian (that is, the mean sun is at maximum elevation to an observer located anywhere on the Greenwich meridian); when the mean sun is 90 deg west of Greenwich, the GMT is 18 h or 6 p.m., and so on. The notation "mod 24" in the preceding equation simply means that integer multiples of 24 are to be added or subtracted as necessary to give GMT a value in the range 0–24 h.

A mean solar day is defined as the time during which the Earth makes a complete rotation relative to the mean sun; the mean sun, of course, moves eastward along the celestial equator at a rate of 360 deg in about 365.25 mean solar days or slightly less than 1 deg/day. The length of a mean solar day is 24 mean solar hours. A sidereal day is the time during which the Earth makes a complete rotation relative to a fixed direction, e.g., the vernal equinox; clearly, it is shorter than a mean solar

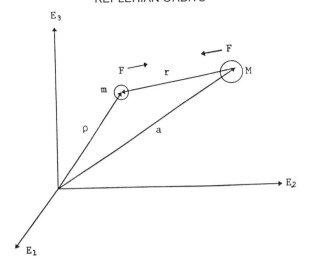

Fig. 3.2 Two-body system.

Therefore,

$$\ddot{r} + \frac{GMr}{r^3} = 0 \qquad (3.8)$$

where $\ddot{r} = \mathrm{d}^2r/\mathrm{d}t^2$ = acceleration of mass m relative to the inertial frame. Equation (3.8) represents the motion of mass m in a gravitational field of mass M, assumed spherically symmetric and, therefore, concentrated at the origin of the reference system.

Equation (3.8) differs from Eq. (3.5) only in the gravitational constant term. The motion of the restricted two-body problem is therefore similar to that of the general two-body system and is affected by the magnitude of the gravitational term. The latter is a negligible effect when $m \ll M$, which is true for the satellites of the Earth and other planetary bodies.

3.3 Conservation of Mechanical Energy

Consider Eq. (3.8) in the form

$$\ddot{r} + \frac{\mu r}{r^3} = 0 \qquad (3.9)$$

where $\mu = GM$, and m is assumed negligibly small compared to M.

Scalar multiplication of Eq. (3.9) by $\dot{r}$ results in

$$\dot{r} \cdot \ddot{r} + \frac{\mu r \cdot \dot{r}}{r^3} = 0 \qquad (3.10)$$

or

$$\frac{\mathrm{d}}{\mathrm{d}t}\left(\frac{\dot{r} \cdot \dot{r}}{2}\right) + \frac{\mu}{r^3}\frac{\mathrm{d}}{\mathrm{d}t}\left(\frac{r \cdot r}{2}\right) = 0 \qquad (3.11)$$

since $d(\mathbf{r} \cdot \mathbf{r})/dt = 2r\dot{r}$, etc. This equation can be integrated to yield

$$\frac{(\dot{r})^2}{2} - \frac{\mu}{r} = \varepsilon$$

$$= \text{specific mechanical energy} \qquad (3.12)$$

Here $(\dot{r})^2/2 = v^2/2 = $ specific kinetic energy and $-\mu/r$ is the specific potential energy of the satellite. The specific potential energy is also equal to the gravitational potential function per unit mass.

3.4   Conservation of Angular Momentum

The specific angular momentum H of a satellite (angular momentum per unit mass) can be obtained by vector-multiplying Eq. (3.9) by $\mathbf{r}$. Then,

$$\mathbf{r} \times \ddot{\mathbf{r}} + \mathbf{r} \times \frac{\mu \mathbf{r}}{r^3} = 0 \qquad (3.13)$$

which shows that

$$\mathbf{r} \times \ddot{\mathbf{r}} = \frac{d}{dt}(\mathbf{r} \times \dot{\mathbf{r}})$$

$$= \frac{d}{dt}\mathbf{H}$$

$$= 0 \qquad (3.14)$$

Consequently, $\mathbf{H} = $ const. This means that $\mathbf{r}$ and $\dot{\mathbf{r}}$ are always in the same plane. The actual solution for the satellite motion can be obtained by cross-multiplying Eq. (3.9) by $\mathbf{H}$. Then,

$$\ddot{\mathbf{r}} \times \mathbf{H} = \frac{\mu}{r^3}(\mathbf{H} \times \mathbf{r}) \qquad (3.15)$$

or

$$\frac{d}{dt}(\dot{\mathbf{r}} \times \mathbf{H}) = \frac{\mu}{r^3}(r^2\dot{\theta})r\hat{\theta}$$

$$= \mu\dot{\theta}\hat{\theta}$$

$$= \mu\frac{d}{dt}(\hat{r}) \qquad (3.16)$$

where the magnitude of the specific angular momentum $H = r^2\dot{\theta}$, $\hat{\theta}$ is a unit vector normal to the unit vector $\hat{r}$ along the $\mathbf{r}$ vector, and $\dot{\theta}$ is the angular rate of the $\mathbf{r}$ vector.

Integrating Eq. (3.16), one obtains

$$\dot{\mathbf{r}} \times \mathbf{H} = \mu\hat{r} + \mathbf{B} \qquad (3.17)$$

where $\mathbf{B}$ is a constant of integration.

Furthermore, since

$$\boldsymbol{r} \cdot (\dot{\boldsymbol{r}} \times \boldsymbol{H}) = \boldsymbol{r} \cdot (\mu \hat{\boldsymbol{r}} + \boldsymbol{B})$$

$$= (\boldsymbol{r} \times \dot{\boldsymbol{r}}) \cdot \boldsymbol{H}$$

$$= \boldsymbol{H} \cdot \boldsymbol{H}$$

$$= H^2$$

$$H^2 = \mu r + r B \cos \theta \qquad (3.18)$$

therefore,

$$r = \frac{H^2/\mu}{1 + (B/\mu)\cos\theta} \qquad (3.19)$$

where

$$H^2/\mu = p = \text{semilatus rectum}$$

$$B/\mu = e = \left(1 + \frac{2\varepsilon H^2}{\mu^2}\right)^{1/2} = \text{eccentricity}$$

$$\theta = \text{true anomaly}$$

The general equation for the radius r in Eq. (3.9) is therefore of the form

$$r = \frac{p}{1 + e \cos\theta} \qquad (3.20)$$

This is an equation for a conic section, an example of which is the ellipse illustrated in Fig. 3.3.

3.5 Orbital Parameters of a Satellite

The orbit ellipse geometry is shown in Fig. 3.3. The following notation is used:

a = semimajor axis = $(r_a + r_p)/2$
b = semiminor axis
e = eccentricity = $(r_a - r_p)/(r_a + r_p)$
θ = true anomaly
r_a = apogee radius = $a(1 + e)$
r_p = perigee radius = $a(1 - e)$
p = $a(1 - e^2) = b^2/a = r_p(1 + e) = r_a(1 - e)$ = semilatus rectum
γ = flight-path angle
  = $\pi/2 - \beta$

The radial velocity component v_r can be found as follows:

$$v_r = \frac{dr}{dt} = \frac{dr}{d\theta}\frac{d\theta}{dt} = \frac{dr}{d\theta}\dot{\theta}$$

$$= \frac{dr}{d\theta} \cdot \frac{H}{r^2} \qquad (3.21)$$

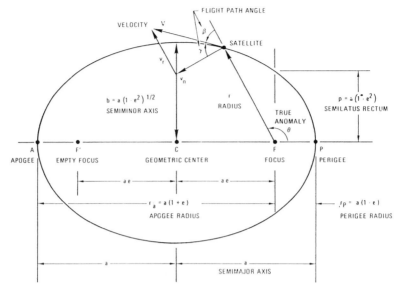

Fig. 3.3   Ellipse geometry (from Ref. 2).

but, since

$$\frac{dr}{d\theta} = \frac{d}{d\theta}[p(1 + e\cos\theta)^{-1}]$$

$$v_r = \frac{He\sin\theta}{p}$$

$$= \sqrt{\frac{\mu}{p}}\,e\sin\theta \tag{3.22}$$

The normal component v_n is found as

$$v_n = r\dot{\theta} = r\left(\frac{H}{r^2}\right) = \sqrt{\frac{\mu}{p}}(1 + e\cos\theta) \tag{3.23}$$

The flight-path angle γ is given by

$$\gamma = \cos^{-1}\left(\frac{v_n}{v}\right) = \tan^{-1}\left(\frac{v_r}{v_n}\right) \tag{3.24}$$

where

$$v = \left(v_n^2 + v_r^2\right)^{1/2}$$

$$= \left[\frac{\mu}{p}(1 + e^2 + 2e\cos\theta)\right]^{1/2} \tag{3.25}$$

4
Position and Velocity as a Function of Time

4.1   General Relationships

In this chapter, we will discuss what is known historically as "Kepler's problem." Succinctly stated, this problem is one of finding the state (position and velocity) of an object in orbit at a specified time t, given the state at some reference time t_0.

For example, let us assume that, at the reference (or initial condition) time, the object is not necessarily at its perifocal point. That is, we will assume the state at t_0 to be available in the following terms:

a = semimajor axis
e = eccentricity
i = inclination
Ω = right ascension of the ascending node
ω = argument of perifocal point
θ = true anomaly (not zero in the present discussion)

To fix ideas firmly, let us review what these terms signify. First, a and e specify the size and shape of the orbit as illustrated in Fig. 4.1. The semiminor axis b is related to the semimajor axis a by the relationship of

$$b = \sqrt{a^2 - c^2}$$

where

$$c = ae$$

or

$$b = a\sqrt{1 - e^2}$$

Second, the orientation of the orbit is specified by i, Ω, and ω, i.e., inclination, right ascension of the ascending node, and argument of perifocus, respectively, as illustrated in Fig. 4.2.

Third, the position of the object is specified by the sixth term, θ, true anomaly, which is measured in the orbit plane from the perifocal axis, as shown in Fig. 4.3. In our discussion, the object is not at perifocus and has a nonzero value for θ_0.

The problem at hand is to find the position θ corresponding to time t, which may be before or after the reference time of t_0. Note that the other terms, $a - \omega$, do not change over this time. Such an orbit is called a "Keplerian" orbit, where the only influence experienced by the object is the gravitational force of the attracting body represented by a spherical potential field. The strength of a spherical field is a function only of the distance from the center of the attracting body.

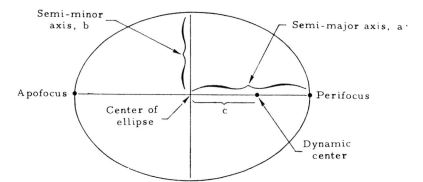

Fig. 4.1   Size and shape of an orbit.

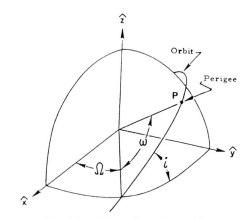

Fig. 4.2   Orientation of an orbit.

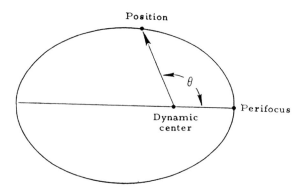

Fig. 4.3   Position of true anomaly.

Generally speaking, the potential field is not spherical, and the first five terms, $a - \omega$, do change as a function of time. There are, however, many problems that can be solved adequately by considering the orbit to be Keplerian. Non-Keplerian influences like the nonspherical potential, atmospheric drag, lunar and planetary effects, and solar pressure are referred to as "perturbative" effects and are discussed more fully in Chapter 8.

Now, let us solve our problem. First, we must relate the initial position θ_0 to the initial time t_0. Equations (4.1–4.3) transform θ_0, the true anomaly at t_0, to its equivalent eccentric anomaly, usually symbolized by the letter E_0.

$$\sin E_0 = \frac{\sqrt{1 - e^2}\,\sin\theta_0}{1 + e\cos\theta_0} \tag{4.1}$$

$$\cos E_0 = \frac{e + \cos\theta_0}{1 + e\cos\theta_0} \tag{4.2}$$

$$E_0 = \tan^{-1}\left(\frac{\sin E_0}{\cos E_0}\right) \tag{4.3}$$

Kepler's equation then relates the eccentric anomaly E_0 to its mean anomaly M_0.

$$M_0 = E_0 - e\sin E_0 \tag{4.4}$$

Finally, the mean anomaly M_0 is related to time t_0 by

$$M_0 = n(t_0 - T) \tag{4.5}$$

where n is the mean motion and T the time at the last previous perifocus passage. We need not determine the exact value of T but merely note its existence for the time being. The mean motion is determined from

$$n = \frac{2\pi}{P} \tag{4.6}$$

where

$$P = 2\pi\sqrt{a^3/\mu} \tag{4.7}$$

In Eq. (4.7), a is the semimajor axis that was given at the start of this problem, and μ is the product of the universal gravitation constant and the mass of the attracting body. If the attracting body is the Earth, μ has the value of $398{,}600.8$ km^3/s^2. You may recognize Eq. (4.7) as Kepler's third law, which states that the square of the orbit period P is proportional to the cube of the orbit's semimajor axis a.

Before proceeding, let us look at these equations from a geometrical view. Figure 4.4 illustrates the relationship of θ and E as expressed in Eqs. (4.1–4.3). For convenience and to demonstrate generality, the subscript zeros have been temporarily removed in the geometrical discussion.

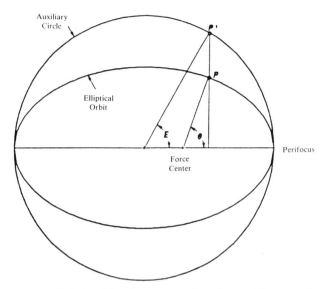

Fig. 4.4   Relationship of true anomaly and eccentric anomaly.

The circle that just fits snugly outside the elliptical orbit is called the "auxiliary circle." It has nothing to do with reality but is a convenient concept introduced to relate position and time. From Kepler's second law, we readily see that, just as the motion of point P in the elliptical orbit is not uniform (i.e., it moves faster when near the perifocus and slower in the region of apofocus), the motion of its image P' in the auxiliary circle is also not uniform.

Now, let us consider another auxiliary circle, as shown in Fig. 4.5. A position P'' on this circle is described by a variable called the mean anomaly M, which is an angular quantity measured at the center of the circle from some reference direction. As we shall see in a moment, the motion of P'' is uniform, i.e., it revolves at a constant speed that is not like the motion of P and P'.

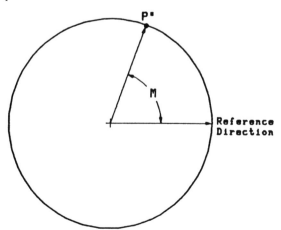

Fig. 4.5   Auxiliary circle of mean anomaly.

Recently, Conway,[6] of the University of Illinois, has applied a root-finding method of Laguerre (1834–1886) to the solution of Kepler's equation. Although the method is intended for finding the roots of a polynomial, it works equally as well for Kepler's equation, which is transcendental. Laguerre's iteration method requires the calculation of $f(E)$, $f'(E)$, and $f''(E)$ at each step and has these remarkable properties:

1) It is cubically convergent for simple roots.
2) For algebraic equations with only real roots, it is convergent for every choice of real initial estimate.

In over 500,000 test solutions, Conway has always found the algorithm to converge to the proper value of eccentric anomaly.

Mathematically, the method consists of solving successively the following equation:

$$E_{i+1} = E_i - \frac{nf(E_i)}{f'(E_i) \pm \sqrt{H(E_i)}} \qquad (4.23)$$

where

$$H(E_i) = \left|(n-1)\left\{(n-1)\left[f'(E_i)\right]^2 - nf(E_i)f''(E_i)\right\}\right| \qquad (4.24)$$

Normally, n is the degree of the polynomial. For our purposes here, we may safely use an arbitrary choice of $n = 5$, making Eqs. (4.23) and (4.24) appear as

$$E_{i+1} = E_i - \frac{5f(E_i)}{f'(E_i) \pm 2\sqrt{\left|4\left[f'(E_i)\right]^2 - 5f(E_i)f''(E_i)\right|}} \qquad (4.25)$$

where

$$f(E_i) = E_i - e \sin E_i - M \qquad (4.26)$$

$$f'(E_i) = 1 - e \cos E_i \qquad (4.27)$$

$$f''(E_i) = e \sin E_i \qquad (4.28)$$

Note that, when Eq. (4.25) is calculated, the sign in the denominator should be chosen so that $|E_{i+1} - E_i|$ is small as possible.

Figures 4.8 and 4.9 show the comparable results when Laguerre's method is used in place of Newton's method. Again, $E_0 = M$ is used as the initial starting value for all cases. Note the remarkable improvement in the number of iterations required, especially in the region where e approaches 1 and M approaches 0. Note also the lack of sensitivity to the starting value. Even with the unsophisticated choice of $E_0 = M$, four iterations at the most are sufficient for Laguerre's method to converge to within 10^{-12} rad almost everywhere in the e-M plane. The use of a good starting value would improve the speed of convergence, but this seems unnecessary.

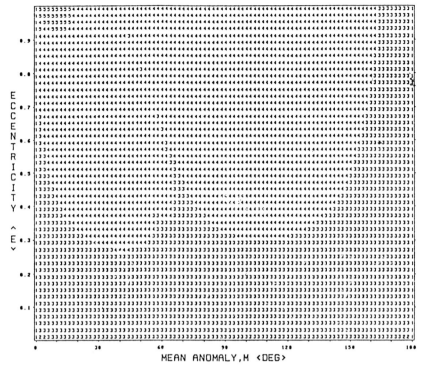

Fig. 4.8   Number of iterations needed to converge by Laguerre's method using the starting value of $E_0 = M$.

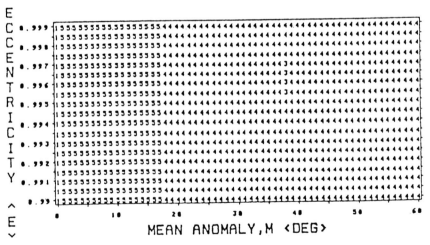

Fig. 4.9   Number of iterations needed to converge by Laguerre's method in the high-e/low-M region using the starting value of $E_0 = M$.

For those who wish to use a starting value other than $E_0 = M$, the following formula is suggested:

$$E_0 = M + \frac{e \sin M}{B + M \sin e} \qquad (4.29)$$

where

$$B = \cos e - \left(\frac{\pi}{2} - e\right) \sin e \qquad (4.30)$$

The computational requirements of this starting value are modest, and the resulting number of iterations needed to solve Kepler's equation is shown in Figs. 4.10 and 4.11.

As a comparison with Figs. 4.8 and 4.9 shows, one iteration less is needed in most instances to arrive at the same accuracy. In some instances, the number of iterations is reduced by two. Gratifyingly, this reduction extends in part to the high-e/low-M region.

It is interesting to note how close the starting value is to the actual value. This closeness can be seen in Figs. 4.12 and 4.13, where the percent error is noted for each starting value in the e-M plane. Specifically, the following percent error is

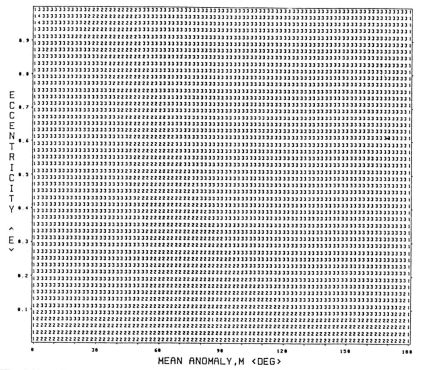

Fig. 4.10   **Number of interations needed to converge by Laguerre's method using the starting value of Eqs. (4.29) and (4.30).**

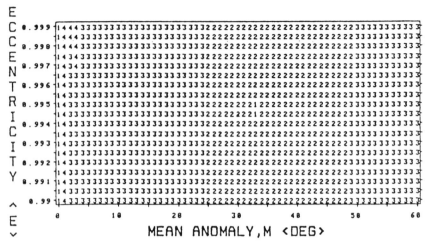

Fig. 4.11   Number of interations needed to converge by Laguerre's method in the high-*e*/low-*M* region using the starting value of Eqs. (4.29) and (4.30)

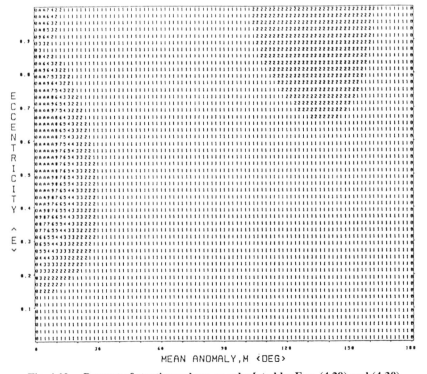

Fig. 4.12   Percent of starting values, as calculated by Eqs. (4.29) and (4.30).

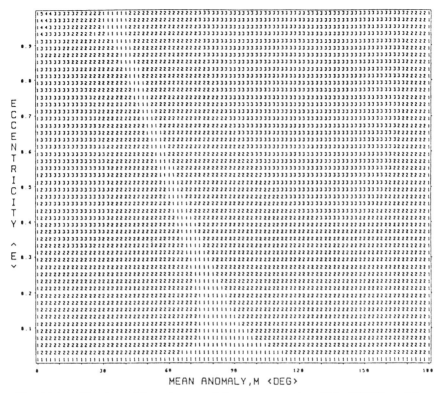

Fig. 4.16c   Number of iterations needed to converge to within 0.00001 rad by Newton's method using the starting value of Eqs. (4.29) and (4.30).

We can carry this analysis one step further by plotting the computing time of Newton's method with starting values of Eqs. (4.29) and (4.30). This is shown as the dot-dash line in Fig. 4.17. Newton's method with $E_0 = M$ is also shown (as a solid line) but is not considered in this analysis because of the diverging behavior observed in Figs. 4.6 and 4.7.

In comparing Laguerre's (long-dash line) and Newton's (dot-dash line) methods, we see that 3 iterations by Newton's method require less computing time than 2 iterations by Laguerre's method. For 3 iterations by Laguerre's method, 4 or 5 iterations by Newton's method are computationally faster. For 4 iterations by Laguerre, 5, 6, or 7 iterations by Newton are faster, and so on. A comparison of Figs. 4.10 and 4.14 can identify the e-M region where Newton's method, with Eqs. (4.29) and (4.30) as starting values, is computationally faster than Laguerre's method, with similar starting values. Again, for ease of comparison, Figs. 4.20 and 4.21 are provided, where the shaded (asterisk) region identifies faster convergence by Newton's method over Laguerre's method.

After these figures are examined, it seems reasonable to adopt Newton's method, with starting values of Eqs. (4.29) and (4.30) for all e-M values except for the region of $e \geq 0.99$, and $M < 4$ deg, where Laguerre's method, with similar starting values, can be applied.

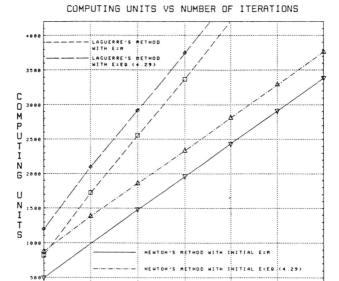

COMPUTING UNITS VS NUMBER OF ITERATIONS

Fig. 4.17   Computing times to solve Kepler's equation.

Now that we know how to solve Kepler's equation analytically (albeit in an iterative fashion), is that all there is to it? Unfortunately, the answer is no. What we have just solved is the situation in which the orbit of concern is elliptical. If the orbit is hyperbolic, we must use Kepler's equation in the form of

$$M = e \sinh F - F \tag{4.32}$$

If the orbit is parabolic, there is yet another form of Kepler's equation, which we will not pursue at this moment.

In the hyperbolic case, it is not enough to change Kepler's equation; the expressions relating the eccentric and true anomalies must also be changed as follows. Equations (4.1–4.3) are changed to

$$\sinh F = \frac{\sqrt{e^2 - 1}\,\sin\theta}{1 + \cos\theta} \tag{4.33}$$

$$\cosh F = \frac{\cos\theta + e}{1 + \cos\theta} \tag{4.34}$$

$$F = \tanh^{-1}\left(\frac{\sinh F}{\cosh F}\right) \tag{4.35}$$

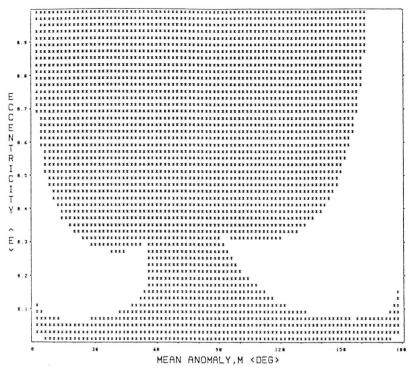

MEAN ANOMALY, M <DEG>

Fig. 4.18   Region (shaded) where Laguerre's method, with starting values of Eqs. (4.29) and (4.30), is computationally faster than the same method with $E_0 = M$.

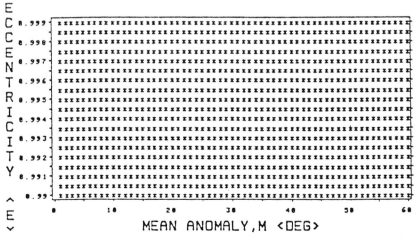

MEAN ANOMALY, M <DEG>

Fig. 4.19   High-e/low-M region (shaded) where Laguerre's method, with starting values of Eqs. (4.29) and (4.30), is computationally faster than the same method with $E_0 = M$.

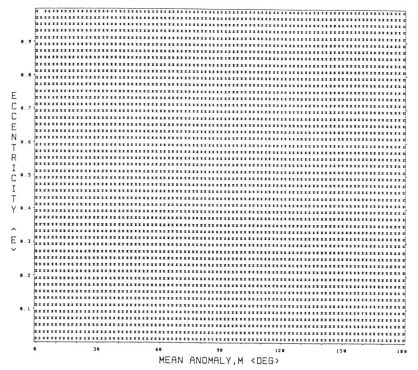

Fig. 4.20   Region (shaded) where Newton's method, with Eqs. (4.29) and (4.30), is computationally faster than Laguerre's method, with Eqs. (4.29) and (4.30).

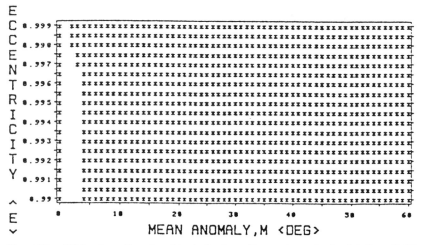

Fig. 4.21   High-*e*/low-*M* region (shaded) where Newton's method, with Eqs. (4.29) and (4.30), is computationally faster than the Laguerre's method with Eqs. (4.29) and (4.30).

Now we may apply these three equations to

$$X_{i+1} = X_i - \frac{5f(X_i)}{f'(X_i) \pm 2\sqrt{|4[f'(X_i)]^2 - 5f(X_i)f''(X_i)|}} \qquad (4.67)$$

where the sign in the denominator is chosen so that $|X_{i+1} - X_i|$ is as small as possible for each iteration.

Note that the t in Eq. (4.64) is, in reality, $\Delta t = t - t_0$. Our choice of $t_0 = 0$ was arbitrary in formulating a solution using the universal variables. Since, in general, $t_0 \neq 0$, a correct Δt must be calculated for the rightmost term of Eq. (4.64). Finally, to start the iterative process, the following initial X value is suggested:

$$X_0 = \sqrt{\mu}\,\frac{t - t_0}{|a|} \qquad (4.68)$$

If the orbit is elliptic, the term $t - t_0$ in the numerator should be reduced by the largest integer multiple of P, where $P = 2\pi a^{3/2}/\sqrt{\mu}$. Other initial approximations may speed the convergence of the solution but seem unnecessary with Laguerre's method of iteration.

4.4 Expressions with *f* and *g*

Once X corresponding to t is determined, we must now find r and V in terms of X, r_0 and V_0.

Since Keplerian motion is confined to a plane, the four vectors r, V, r_0, and V_0 are all coplanar. Thus, we can write

$$r = fr_0 + gV_0 \qquad (4.69)$$

and differentiating

$$V = \dot{f}r_0 + \dot{g}V_0 \qquad (4.70)$$

where f, g, $\dot{f}$ and $\dot{g}$ are time-dependent scalars.

One interesting property of the f and g terms is seen by crossing r and V as follows:

$$r \times V = h = (f\dot{g} - \dot{f}g)h \qquad (4.71)$$

from which

$$1 = f\dot{g} - \dot{f}g \qquad (4.72)$$

This relationship states that f, g $\dot{f}$, and $\dot{g}$ are not independent and, if we know any three, we can determine the fourth from this identity.

Now, the approach taken to develop these f and g expressions is 1) to write Eqs. (4.69) and (4.70) in terms of a perifocal coordinate system, 2) relate Eq. (4.46) to the conic equation and find expressions of the perifocal components in terms of X, 3) substitute the results of step 2 into the results of step 1, and 4) introduce the definitions of Z, $S(Z)$, and $C(Z)$. We shall omit the details of these steps, but the results of such an endeavor are

$$f = 1 - \frac{X^2}{r_0}C \tag{4.73}$$

$$g = t - \frac{X^3}{\sqrt{\mu}} \tag{4.74}$$

$$\dot{f} = \frac{\sqrt{\mu}}{rr_0}X(ZS - 1) \tag{4.75}$$

$$\dot{g} = 1 - \frac{X^2C}{r} \tag{4.76}$$

4.5 Summary of the Universal Approach

As a summary of the universal approach discussed so far, we will outline the entire procedure and pertinent equations as follows:

Given: r_0 and V_0 at t_0 and t

$$r_0 = |\mathbf{r}_0| \tag{4.77}$$

$$V_0 = |\mathbf{V}_0| \tag{4.78}$$

$$\frac{1}{a} = \frac{2\mu/r_0 - V_0^2}{\mu} \tag{4.79}$$

$$\Delta t = t - t_0 \tag{4.80}$$

If $\frac{1}{a} \leq 0$, skip to Eq. (4.83). Otherwise,

$$P = 2\pi\frac{a^{3/2}}{\sqrt{\mu}} \tag{4.81}$$

$$\Delta t = \Delta t - [\text{sign}(\Delta t)]\text{int}\left[\frac{|\Delta t|}{P}\right]P \tag{4.82}$$

where int $|\xi|$ denotes the integer part of ξ.

$$X_0 = \sqrt{\mu}\frac{\Delta t}{|a|} \tag{4.83}$$

Start with $n = 0$, and calculate

$$Z_n = \frac{X_n^2}{a} \tag{4.84}$$

$$C_n = \frac{1}{2!} - \frac{Z_n}{4!} + \frac{Z_n^2}{6!} - \frac{Z_n^3}{8!} - \cdots \tag{4.85}$$

$$S_n = \frac{1}{3!} - \frac{Z_n}{5!} + \frac{Z_n^2}{7!} - \frac{Z_n^3}{9!} + \cdots \tag{4.86}$$

$$f(X_n) = \left(1 - \frac{r_0}{a}\right) S_n X_n^3 + \frac{r_0 \cdot V_0}{\sqrt{\mu}} C_n X_n^2 + r_0 X_n - \sqrt{\mu} t \tag{4.87}$$

$$f'(X_n) = C_n X_n^2 + \frac{r_0 \cdot V_0}{\sqrt{\mu}} (1 - S_n Z_n) X_n + r_0 (1 - C_n Z_n) \tag{4.88}$$

$$f''(X_n) = \left(1 - \frac{r_0}{a}\right)(1 - S_n Z_n) X_n + \frac{r_0 \cdot V_0}{\sqrt{\mu}} (1 - C_n Z_n) \tag{4.89}$$

$$\delta_n = 2\sqrt{4[f'(X_n)]^2 - 5f(X_n)f''(X_n)} \tag{4.90}$$

$$\Delta X_n = \frac{5f(X_n)}{f'(X_n) + [\text{sign } f'(X_n)]\delta_n} \tag{4.91}$$

$$X_{n+1} = X_n - \Delta X_n \tag{4.92}$$

Repeat Eqs. (4.84–4.92) for $n = 1, 2, 3 \ldots$ until

$$\left| \frac{(\Delta X_n)^2}{a} \right| < \varepsilon \tag{4.93}$$

where $\varepsilon = 10^{-8}$.

$$f = 1 - \frac{X^2}{r_0} C \tag{4.94}$$

$$g = t - \frac{X^3}{\sqrt{\mu}} S \tag{4.95}$$

$$\dot{f} = \frac{\sqrt{\mu}}{r r_0} (SZ - 1) X \tag{4.96}$$

$$\dot{g} = 1 - \frac{X^2}{r} C \tag{4.97}$$

$$r = f r_0 + g V_0 \tag{4.98}$$

$$V = \dot{f} r_0 + \dot{g} V_0 \tag{4.99}$$

4.6 The Classical Element Set

Let us pause for the moment and examine the coordinate frames we have used so far. When we first described Kepler's problem, we started with an element set that consisted of the following: a, e, i, Ω, ω, and θ. This is a modified form of the classical element set in which the sixth element is θ (true anomaly). Sometimes, M (mean anomaly) is used instead of θ. Rarely is E (eccentric anomaly) used. The true classical element set consists of a, e, i, Ω, ω, and τ, where the sixth term, τ, is the time of perifocal passage or, more exactly, the time of the last previous perifocal passage as measured from a specified reference time (usually midnight of Greenwich mean time of an epoch date). It is because of the difficulty encountered in handling this sixth term that practical considerations have led to the use of θ and M in its place.

4.7   The Rectangular Coordinate System

When we discussed the universal variables, the initial and final orbit states were expressed by two quantities, position and velocity, in their vector form. Position and velocity as vectors can be resolved into a variety of components. Most common are those expressed in a rectangular Cartesian coordinate frame, in which case the components are x, y, z, $\dot{x}$, $\dot{y}$, and $\dot{z}$. Often, this coordinate frame is referred to as the Earth-centered inertial (ECI) frame.

Since it is apparent that we need to change from one coordinate frame to another, this is a good time to see how we transform from the modified classical element set to the rectangular Cartesian set and vice versa.

4.8   Modified Classical to Cartesian Transformation

This transformation is accomplished in two steps. First, the classical set is expressed in a perifocal coordinate frame. Second, the perifocal coordinate frame is converted to a Cartesian frame through a series of rotations of the axes.

Looking at Fig. 4.22, we can immediately write

$$\boldsymbol{r} = r \cos\theta \hat{\boldsymbol{P}} + r \sin\theta \hat{\boldsymbol{Q}} \tag{4.100}$$

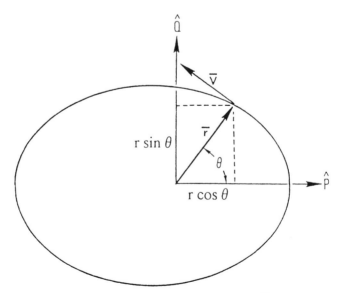

Fig. 4.22   Position and velocity in a perifocal coordinate system.

where the magnitude of r is determined from the equation of the conic

$$r = \frac{p}{1 + e \cos \theta} \tag{4.100a}$$

p being equal to the semilatus rectum of the orbit, which is $a(1 - e^2)$. Differentiating and simplifying the preceding two equations yields

$$V = \sqrt{\frac{\mu}{p}}[(-\sin \theta)\hat{P} + (e + \cos \theta)\hat{Q}] \tag{4.101}$$

In differentiating, note that $\dot{\hat{P}} = \dot{\hat{Q}} = 0$ since the perifocal coordinate frame is "inertial" in space. Also, $\dot{r} = \sqrt{(\mu/p)}e \sin \theta$ and $r\dot{\theta} = \sqrt{(\mu/p)}(1 + e \cos \theta)$ from Eq. (4.41) and Eq. (4.100a) and its derivative.

Now, from Fig. 4.23, we see that the IJK axes can become the PQW axes by three successive rotations as follows: 1) rotation about the $\hat{z}$ axis by $+\Omega$, 2) rotation about the $\hat{x}$ axis by $+i$, and 3) rotation about the $\hat{z}$ axis by $+\omega$. The first transformation is accomplished by

$$\begin{bmatrix} x' \\ y' \\ z' \end{bmatrix} = \begin{bmatrix} \cos \Omega & \sin \Omega & 0 \\ -\sin \Omega & \cos \Omega & 0 \\ 0 & 0 & 1 \end{bmatrix} \begin{bmatrix} x \\ y \\ z \end{bmatrix}_{IJK} \tag{4.102}$$

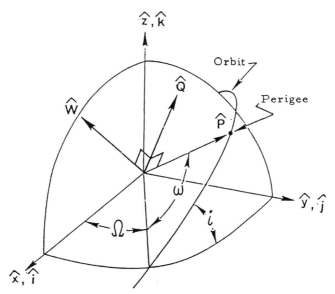

Fig. 4.23   The relationship between the IJK and PQW systems.

The second transformation is accomplished by

$$
\begin{bmatrix} x'' \\ y'' \\ z'' \end{bmatrix} = \begin{bmatrix} 1 & 0 & 0 \\ 0 & \cos i & \sin i \\ 0 & -\sin i & \cos i \end{bmatrix} \begin{bmatrix} x' \\ y' \\ z' \end{bmatrix} \tag{4.103}
$$

and the third transformation by

$$
\begin{bmatrix} x''' \\ y''' \\ z''' \end{bmatrix}_{PQW} = \begin{bmatrix} \cos \omega & \sin \omega & 0 \\ -\sin \omega & \cos \omega & 0 \\ 0 & 0 & 1 \end{bmatrix} \begin{bmatrix} x'' \\ y'' \\ z'' \end{bmatrix} \tag{4.104}
$$

Actually, we need to change the *PQW* axes to the *IJK* axes. That is, instead of Eq. (4.102), we need

$$
\begin{bmatrix} x \\ y \\ z \end{bmatrix}_{IJK} = [T_1] \begin{bmatrix} x' \\ y' \\ z' \end{bmatrix} \tag{4.105}
$$

where $[T_1]$ is the inverse of the transformation matrix given in Eq. (4.102). Similarly, we need

$$
\begin{bmatrix} x' \\ y' \\ z' \end{bmatrix} = [T_2] \begin{bmatrix} x'' \\ y'' \\ z'' \end{bmatrix} \quad \text{and} \quad \begin{bmatrix} x'' \\ y'' \\ z'' \end{bmatrix} = [T_3] \begin{bmatrix} x''' \\ y''' \\ z''' \end{bmatrix}_{PQW} \tag{4.106}
$$

where $[T_2]$ and $[T_3]$ are the inverses of the transformation matrices given in Eqs. (4.103) and (4.104). The inverses are obtained simply by changing the sign to the sine terms of the matrices. Thus, we have

$$
[T_1] = \begin{bmatrix} \cos \Omega & -\sin \Omega & 0 \\ \sin \Omega & \cos \Omega & 0 \\ 0 & 0 & 1 \end{bmatrix} \tag{4.107}
$$

$$
[T_2] = \begin{bmatrix} 1 & 0 & 0 \\ 0 & \cos i & -\sin i \\ 0 & \sin i & \cos i \end{bmatrix} \tag{4.108}
$$

and

$$
[T_3] = \begin{bmatrix} \cos \omega & -\sin \omega & 0 \\ \sin \omega & \cos \omega & 0 \\ 0 & 0 & 1 \end{bmatrix} \tag{4.109}
$$

The complete transformation is then accomplished by three successive transformations

$$\begin{bmatrix} x \\ y \\ z \end{bmatrix}_{IJK} = [T_1][T_2][T_3] \begin{bmatrix} x \\ y \\ z \end{bmatrix}_{PQW} \tag{4.110}$$

Combining the three transformation matrices into one then gives

$$\begin{bmatrix} x \\ y \\ z \end{bmatrix} = [R] \begin{bmatrix} r\cos\theta \\ r\sin\theta \\ 0 \end{bmatrix} \tag{4.111}$$

and

$$\begin{bmatrix} \dot{x} \\ \dot{y} \\ \dot{z} \end{bmatrix} = [R] \begin{bmatrix} -\sqrt{\frac{\mu}{p}}\sin\theta \\ \sqrt{\frac{\mu}{p}}(e + \cos\theta) \\ 0 \end{bmatrix} \tag{4.112}$$

where

$$[R] = \begin{bmatrix} R_{11} & R_{12} & R_{13} \\ R_{21} & R_{22} & R_{23} \\ R_{31} & R_{32} & R_{33} \end{bmatrix} \tag{4.113}$$

and

$$
\begin{aligned}
R_{11} &= \cos\Omega\cos\omega - \sin\Omega\sin\omega\cos i \\
R_{12} &= -\cos\Omega\sin\omega - \sin\Omega\cos\omega\cos i \\
R_{13} &= \sin\Omega\sin i \\
R_{21} &= \sin\Omega\cos\omega + \cos\Omega\sin\omega\cos i \\
R_{22} &= -\sin\Omega\sin\omega + \cos\Omega\cos\omega\cos i \\
R_{23} &= -\cos\Omega\sin i \\
R_{31} &= \sin\omega\sin i \\
R_{32} &= \cos\omega\sin i \\
R_{33} &= \cos i
\end{aligned} \tag{4.114}
$$

4.9 Rectangular to Modified Classical Elements Transformation

In this transformation, we start with $r(x, y, z)$ and $V(\dot{x}, \dot{y}, \dot{z})$ and seek to change these into the set of $(a, e, i, \Omega, \omega$ and $\theta)$. The computational steps for this transformation are as follows:

$$r = \sqrt{x^2 + y^2 + z^2} \tag{4.115}$$

$$V = \sqrt{\dot{x}^2 + \dot{y}^2 + \dot{z}^2} \tag{4.116}$$

$$\frac{V^2}{\mu} = \frac{2}{r} - \frac{1}{a} \to a \tag{4.117}$$

$$\hat{W} = \frac{r \times V}{|r \times V|} \tag{4.118}$$

$$\cos i = \hat{W} \cdot \hat{k} \to i \tag{4.119}$$

$$e = \frac{1}{\mu} \left[\left(V^2 - \frac{\mu}{r} \right) r - (r \cdot V)V \right] \to e \tag{4.120}$$

$$\hat{N} = \frac{\hat{k} \times \hat{W}}{|\hat{k} \times \hat{W}|} \tag{4.121}$$

$$\cos \Omega = \hat{i} \cdot \hat{N} \tag{4.122}$$

$$\sin \Omega = (\hat{i} \times \hat{N}) \cdot \hat{k} \tag{4.123}$$

$$\Omega = \tan^{-1} \left(\frac{\sin \Omega}{\cos \Omega} \right) \tag{4.124}$$

$$\cos \omega = \frac{\hat{N} \cdot e}{|e|} \tag{4.125}$$

$$\sin \omega = \frac{\hat{N} \times e}{|e|} \cdot \hat{W} \tag{4.126}$$

$$\omega = \tan^{-1} \left(\frac{\sin \omega}{\cos \omega} \right) \tag{4.127}$$

$$\cos \theta = \frac{e \cdot r}{|e||r|} \tag{4.128}$$

$$\sin \theta = \frac{e \times r}{|e||r|} \cdot \hat{W} \tag{4.129}$$

$$\theta = \tan^{-1} \left(\frac{\sin \theta}{\cos \theta} \right) \tag{4.130}$$

4.10   The Spherical (ADBARV) Coordinate System

Another often used and practical coordinate system is the spherical coordinate system. In this system, the position and velocity are expressed in terms of the following six quantities:

α = right ascension
δ = declination
β = flight-path angle
A = azimuth
r = radius
V = velocity

All of these are scalar quantities and, because of the symbols used, the coordinate system is often referred to as the ADBARV system. The first two quantities reflect the fact that the spherical system is similar to the celestial (right ascension-declination) system used in astronomy.

Figure 4.24 is a pictorial representation of the ADBARV components in the rectangular x, y, z coordinate system.

In the Fig. 4.24, α, δ, and the magnitudes of r and V are self-evident.

The flight-path angle β is the angle between the radius and velocity vectors. At either perigee or apogee, the velocity vector is perpendicular to the radius vector, and $\beta = 90$ deg. On half of the orbit from perigee to apogee, β is less than 90 deg. On the other half, from apogee to perigee, β is greater than 90 deg. Depending

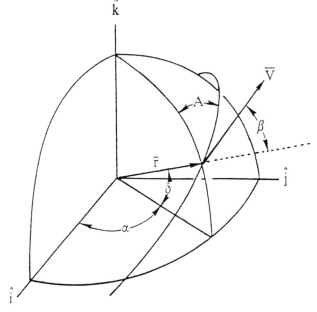

Fig. 4.24   The spherical coordinate system.

on the discipline to which one belongs (control, guidance, etc.), sometimes the flight-path angle is measured relative to the instantaneous geocentric horizontal rather than the vertical. In this case, $\gamma = 90$ deg $-\beta$, and the flight-path angle γ is zero at perigee and apogee. It is positive when approaching apogee from perigee and negative when approaching perigee from apogee. You may also hear the term "pitch angle" used to describe the flight-path angle γ. Here, nose up and nose down are equivalent to the positive and negative sense of the angle, γ.

The azimuth A is measured in the instantaneous geocentric horizontal plane at the point in question. It is the angle between the northerly direction and the projection of the velocity vector onto this plane. Typically, it is positive when measured clockwise from due north when viewed down along the radius vector toward the center of the Earth. Care should be taken in noting the positive/negative sense of this component since some disciplines define the counterclockwise direction as being the positive direction of this angle.

In referring to the instantaneous horizontal, the term "geocentric" has been introduced. This means that the horizontal line or plane is perpendicular to the geocentric radius at that instant. There is also a "geodetic" horizontal, which should not be confused with the geocentric. More on this topic is discussed in Sec. 4.14.

4.11 Rectangular to Spherical Transformation

Now, let us see how we can transform to the spherical (ADBARV) coordinate frame from the rectangular Cartesian coordinate system. The computational steps are

$$r = \sqrt{x^2 + y^2 + z^2} \tag{4.131}$$

$$V = \sqrt{\dot{x}^2 + \dot{y}^2 + \dot{z}^2} \tag{4.132}$$

$$\sin \alpha = \frac{y}{\sqrt{x^2 + y^2}} \tag{4.133}$$

$$\cos \alpha = \frac{x}{\sqrt{x^2 + y^2}} \tag{4.134}$$

$$\alpha = \tan^{-1}\left(\frac{\sin \alpha}{\cos \alpha}\right) \tag{4.135}$$

$$\delta = \sin^{-1}\left(\frac{z}{r}\right) \tag{4.136}$$

$$\beta = \cos^{-1}\left(\frac{r \cdot V}{|r \cdot V|}\right) \tag{4.137}$$

$$\hat{W} = \frac{r \times V}{|r \times V|} \tag{4.138}$$

$$\hat{A} = \frac{\hat{W} \times r}{|\hat{W} \times r|} \tag{4.139}$$

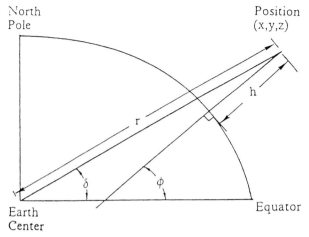

Fig. 4.26   A two-dimensional view of geocentric and geodetic altitudes.

Once these two quantities are determined, the following two equations are used directly.

$$h = r - a_e \left[1 - f \sin^2 \delta - \frac{f^2}{2} \sin^2 2\delta \left(\frac{a_e}{r} - \frac{1}{4} \right) \right] \qquad (4.164)$$

where

$$h = \text{geodetic altitude}$$

$$a_e = \text{equatorial radius of the Earth}$$

$$f = \text{flattening of the Earth}$$

and

$$\sin(\phi - \delta) = \frac{a_e}{r} \left[f \sin 2\delta + f^2 \sin 4\delta \left(\frac{a_e}{r} - \frac{1}{4} \right) \right] \qquad (4.165)$$

From Eq. (4.165), then,

$$\phi = \delta + \sin^{-1} [\sin(\phi - \delta)] \qquad (4.166)$$

where

$$\phi = \text{geodetic latitude of the sublatitude point}$$

The longitude λ of the object is calculated by first determining the right ascension of the position (x, y, z) and then relating it to an Earth-relative frame whose x axis is in the plane of the Greenwich meridian. Figures 4.27 and 4.28 describe the geometry relating to calculating the longitude.

First, from Fig. 4.27,

$$\alpha = \tan^{-1} \left(\frac{y}{x} \right) \qquad (4.167)$$

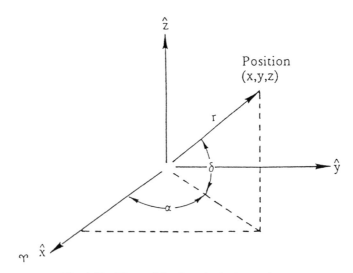

Fig. 4.27   The position in spherical coordinates.

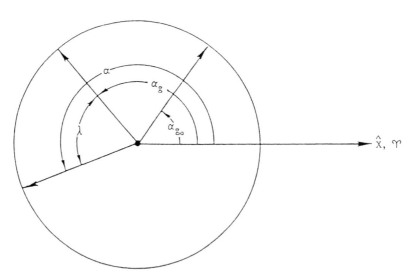

Fig. 4.28   Inertial and Earth-relative angles: α_{g0} = right ascension of Greenwich at epoch; α_g = right ascension of Greenwich at $(t - \text{epoch})$; α = right ascension of the vehicle at $(t - \text{epoch})$.

And, from Fig. 4.28,

$$\lambda = \alpha - \alpha_g \tag{4.168}$$

or

$$\lambda = \{\alpha - [\alpha_{g_0} + \omega_e(t - t_0)]\}_{\text{mod } 360 \text{ deg}}, \qquad 0 \le \lambda < 360 \text{ deg} \tag{4.169}$$

where

$$\omega_e = \text{the rotational rate of the Earth}$$

$$t_0 = \text{the epoch or the reference time}$$

$$\alpha_{g_0} = \text{the right ascension of Greenwich at } t_0$$

By letting t_0 occur at midnight, $\alpha_{g_0} = \alpha_{g@\text{midnight}}$, which can be calculated as shown in Sec. 4.13.

To complete this topic, we will also examine the inverse process, namely, to convert geodetic sublatitude, longitude, and geodetic altitude to the equivalent (x, y, z) position. Unlike earlier conversions, this process is direct and exact and does not use any approximate formulas.

First, as an intermediate step, the geocentric latitude and the geocentric radius of the sublatitude point are determined. Referring to Fig. 4.29, we obtain

$$\phi' = \tan^{-1}[(1 - f)^2 \tan \phi], \qquad -90 \text{ deg} \le \phi \le +90 \text{ deg} \tag{4.170}$$

$$r_E = \frac{a_e(1 - f)}{\sqrt{1 - f(2 - f)\cos^2 \phi'}} \tag{4.171}$$

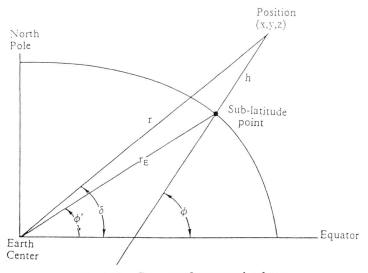

Fig. 4.29   Geometry for converting h to r.

where ϕ' and r_E are the geocentric latitude and radius of the sublatitude point, respectively.

Then

$$x' = r_E \cos \phi' + h \cos \phi \tag{4.172}$$

$$y' = 0 \tag{4.173}$$

$$z' = r_E \sin \phi' + h \sin \phi \tag{4.174}$$

and

$$r = \sqrt{x'^2 + z'^2} \tag{4.175}$$

$$\delta = \sin^{-1}\left(\frac{z'}{r}\right) \tag{4.176}$$

Now, from Eq. (4.169),

$$\alpha = [\lambda + \alpha_{g_0} + \omega_e(t - t_0)]_{\text{mod } 360 \text{ deg}}, \qquad 0 \leq \alpha \leq 360 \text{ deg} \tag{4.177}$$

where, again, by letting t_0 to be at midnight, $\alpha_{g_0} = \alpha_{g@\text{midnight}}$, which can be calculated as shown in Sec. 4.13.

Finally,

$$x = x' \cos \alpha \tag{4.178}$$

$$y = x' \sin \alpha \tag{4.179}$$

$$z = z' \tag{4.180}$$

4.15 Converting from Perigee/Apogee Radii to Perigee/Apogee Altitudes

An interesting variation on what has been described so far is the process to convert a set of perigee/apogee radii to its equivalent set of perigee/apogee altitudes. To be more specific, we have as initial quantities the values for r_p (perigee radius), r_A (apogee radius), and δ_p (declination of the perigee). Inherent in this statement is the assumption that perigee and apogee lie on a straight line that passes through the center of the Earth (i.e., they are 180 deg apart in Earth-centered angle). It follows then that the declination of the apogee is equal in magnitude to the declination of perigee except for its sign, which is opposite. What may not be obvious is that the geodetic sublatitude at perigee is not the same in magnitude as the sublatitude at apogee. Figure 4.30 illustrates this point.

Since r_p and r_A lie on a straight line, $|\delta_p| = |\delta_A|$. But, because $r_p < r_A$, the subapsidal points do not occur at the same latitude (magnitudewise) as shown in the Fig. 4.30. Thus, $|\phi_p| \neq |\phi_A|$. Needless to say, this affects the values of h_p and h_A.

Mathematically, we solve this problem by first using r_p and δ_p in Eqs. (4.164–4.166)

$$h_p = r_p - a_e\left[1 - f \sin^2 \delta_p - \frac{f^2}{2} \sin^2 2\delta_p \left(\frac{a_e}{r_p} - \frac{1}{4}\right)\right] \tag{4.181}$$

$$\phi_p = \delta_p + \sin^{-1}\left\{\frac{a_e}{r_p}\left[f \sin 2\delta_p + f^2 \sin 4\delta_p \left(\frac{a_e}{r_p} - \frac{1}{4}\right)\right]\right\} \tag{4.182}$$

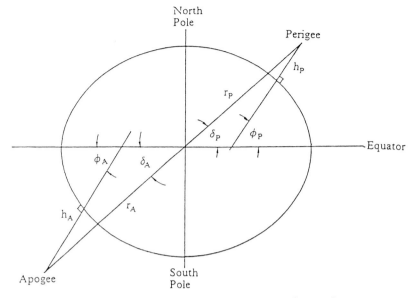

Fig. 4.30   **Differences in sublatitudes between perigee and apogee.**

This process is then repeated using r_A and δ_A, where $\delta_A = -\delta_p$

$$h_A = r_A - a_e \left[1 - f \sin^2 \delta_A - \frac{f^2}{2} \sin^2 2\delta_A \left(\frac{a_e}{r_A} - \frac{1}{4} \right) \right] \qquad (4.183)$$

$$\phi_A = \delta_A + \sin^{-1} \left\{ \frac{a_e}{r_A} \left[f \sin 2\delta_A + f^2 \sin 4\delta_A \left(\frac{a_e}{r_A} - \frac{1}{4} \right) \right] \right\} \qquad (4.184)$$

4.16 Converting from Perigee/Apogee Altitudes to Perigee/Apogee Radii

The most interesting and perhaps most misused conversion occurs when perigee and apogee altitudes are converted to perigee and apogee radii. This occurs typically when perigee and apogee altitudes are converted to their ECI (Earth-centered inertial) counterparts. Symbolically, this conversion is represented by

$$
\text{Orbital elements} \qquad\qquad \text{ECI elements}
$$

$$
\begin{bmatrix} h_p \\ h_A \\ i \\ \phi_{PD} \\ \dot{z}_p/|\dot{z}_p| \\ \lambda_P \end{bmatrix}
\Rightarrow
\begin{bmatrix} x \\ y \\ z \\ \dot{x} \\ \dot{y} \\ \dot{z} \end{bmatrix}
\qquad (4.185)
$$

where

$$i = \text{inclination}$$

$$\phi_{PD} = \text{geodetic sublatitude of perigee}$$

$$\dot{z}_p/|\dot{z}_p| = \text{direction of motion at perigee}$$

$$(+1 \text{ for northbound}; -1 \text{ for southbound})$$

$$\lambda_P = \text{perigee longitude}$$

What is not explicitly stated is that apogee altitude h_A is measured above a geodetic latitude whose magnitude is not equal to the magnitude of the latitude of the subperigee point. Whatever the difference in the magnitudes of the geodetic latitudes, apogee altitude h_A is placed so that the resulting apogee will lie on a straight line that contains both perigee and the center of the Earth.

Mathematically, the following process is used, provided that $h_A \neq 0$ and $h_P < h_A$. First, the perigee radii r_p and its declination δ_P are determined from

$$\phi'_P = \tan^{-1}\left[(1 - f)^2 \tan \phi_{PD}\right] \tag{4.186}$$

$$r_E = \frac{a_E(1 - f)}{\sqrt{1 - f(2 - f)\cos^2 \phi'_P}} \tag{4.187}$$

$$x' = r_E \cos \phi'_p + h_P \cos \phi_{PD} \tag{4.188}$$

$$z' = r_E \sin \phi'_p + h_P \sin \phi_{PD} \tag{4.189}$$

$$r_p = \sqrt{x'^2 + z'^2} \tag{4.190}$$

$$\delta_P = \sin^{-1}\left(\frac{z'}{r}\right) \tag{4.191}$$

Then, the apogee radius r_A and declination δ_A are calculated from

$$r_A = a_e \frac{B + \sqrt{B^2 - C}}{2} \tag{4.192}$$

where

$$B = \left(1 - f \sin^2 \delta_A + \frac{f^2}{8} \sin^2 2\delta_A\right) + \frac{h_A}{a_e} \tag{4.193}$$

$$C = 2f^2 \sin^2 2\delta_A \tag{4.194}$$

and

$$\delta_A = -\delta_P \tag{4.195}$$

Equation (4.192) is derived from Eq. (4.164) by moving h to the right side of the

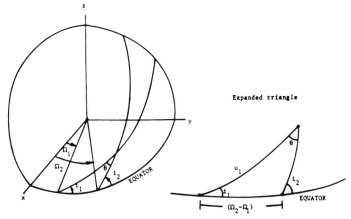

Fig. 5.8   Plane change angle θ in terms of i and Ω

or

$$\frac{\Delta V}{V_{c1}} = 2\sin\theta/2$$

Figure 5.6 plots $\Delta V / V_{c1}$ as a function of θ and shows that plane changes require large ΔV for even modest values of θ. For $\theta = 60$ deg, $\Delta V = V_{c1}$.

Figure 5.8 displays the geometry of a plane change θ, which will, in general, change both the inclination i and the right ascension of ascending node Ω of the original orbit. The plane change will be constrained to an inclination change, i.e., no change in Ω, only if the plane change maneuver is performed at an equatorial crossing. In general, the angle θ between two planes is a function of both inclination i and right ascension of ascending node Ω. Given the initial orbit elements i_1 and Ω_1, the plane change angle θ, and the argument of latitude u_1 of the plane change maneuver, the following two equations from spherical trigonometry can be solved for the final orbit elements, i_2 and Ω_2:

$$\cos i_2 = \cos i_1 \cos\theta - \sin i_1 \sin\theta \cos u_1$$

$$\cos\theta = \cos i_1 \cos i_2 + \sin i_1 \sin i_2 \cos(\Omega_2 - \Omega_1)$$

Another common single-impulse maneuver is a tangential ΔV applied to a circular orbit or to an elliptical orbit at perigee. Figure 5.9 shows a ΔV being added in the direction of motion, i.e., along the velocity vector, to the circular orbit velocity V_c. If the ΔV is relatively small, the resulting orbit is an ellipse, with perigee at the point of ΔV application and apogee located 180 deg away in central angle, i.e., true anomaly, $v = 180$ deg. Now, if the ΔV had been larger or if another relatively small ΔV is added tangentially at the next perigee passage, then the resulting orbit is a larger ellipse with the same perigee but a higher apogee. If the added $\Delta V = V_c(\sqrt{2} - 1)$, then the total velocity $= \sqrt{2}V_c = V_{esc}$ and the resulting orbit is parabolic. And if the added ΔV is greater than this value, then the resulting orbit is hyperbolic. It is interesting to note that, from a circular orbit,

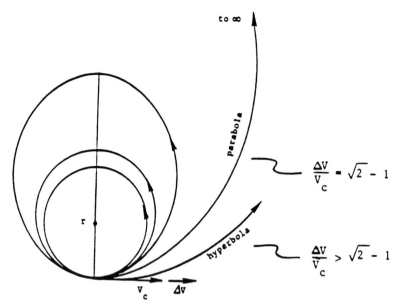

Fig. 5.9 Orbits resulting from a tangential velocity addition.

considerably more ΔV is required to transfer to a rectilinear orbit (drop into the center along a straight line) than to escape.

5.3 Single- and Two-Impulse Transfer Comparison for Coplanar Transfers Between Elliptic Orbits That Differ Only in Their Apsidal Orientation

Single- and Two-Impulse Transfers

For coplanar orbits 1 and 2 in Fig. 5.10, $a_1 = a_2$ and $e_1 = e_2$, but their lines of apsides are rotated by $\Delta \omega$. For single-impulse transfer at either intersection point,

$$\frac{\Delta V/2}{V} = \sin \gamma$$

where γ is the flight-path angle. To solve, substitute the orbit equation $r = p/(1 + e \cos v)$ into the energy equation

$$V^2 = \mu \left[\frac{2}{r} - \frac{1}{a} \right]$$

$$V^2 = \mu \left[\frac{2(1 + e \cos v)}{a(1 - e^2)} - \frac{1}{a} \right] = \mu \left[\frac{2(1 + e \cos v) - (1 - e^2)}{a(1 - e^2)} \right]$$

$$V^2 = \frac{\mu}{p} [2 + 2e \cos v - 1 + e^2] = V^2_{c_{r=p}} [1 + 2e \cos v + e^2]$$

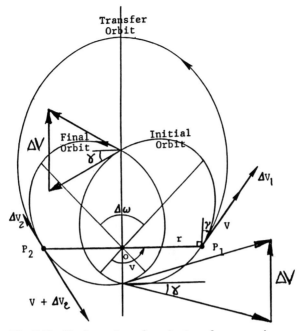

Fig. 5.10   Single- and two-impulse transfer comparison.

and substitute both into the equation for angular momentum, $h = rV \cos \gamma = \sqrt{\mu p}$, squared,

$$\frac{p^2}{(1 + e \cos v)^2} \frac{\mu}{p}[1 + 2e \cos v + e^2] \cos^2 \gamma = \mu p$$

so that

$$\cos \gamma = \frac{1 + e \cos v}{\sqrt{1 + 2e \cos v + e^2}}, \qquad \sin \gamma = \frac{e \sin v}{\sqrt{1 + 2e \cos v + e^2}}$$

Substituting into the ΔV equation,

$$\frac{\Delta V}{V_{c_{r=p}}} = 2e \sin v$$

At the higher intersection,

$$v = 180 \deg \pm \frac{\Delta \omega}{2}$$

and so

$$\frac{\Delta V}{V_{c_{r=p}}} = 2e \sin \frac{\Delta \omega}{2}$$

The solution is exactly the same at the lower intersection.

The solution for the optimal two-impulse cotangential transfer is

$$\frac{\Delta V_{\text{TOTAL}}}{V_{C_{r=p}}} = e \sin \frac{\Delta \omega}{2}$$

when e is assumed small. The sum of the two impulses is half the single-impulse value. The two impulses are equal in magnitude, but one is in the direction of motion while the other is opposite to the direction of motion. The optimal point of application for ΔV_1 is at $v_1 = 90$ deg $+\Delta \omega/2$, and ΔV_2 is applied 180 deg away (see Fig. 5.10). The optimal two-impulse transfer between these orbits is given by Lawden.[1]

5.4 Hohmann Transfer

The Hohmann transfer[2] is the minimum two-impulse transfer between coplanar circular orbits. Derivations of the velocity requirements ΔV_1 and ΔV_2 and the transfer time, as well as a figure of the transfer and plotted results, are presented in the following pages.

Referring to Fig. 5.11, the Hohmann transfer is a relatively simple maneuver. A tangential ΔV_1 is applied to the circular orbit velocity. The magnitude of ΔV_1 is determined by the requirement that the apogee radius of the resulting transfer ellipse must equal the radius of the final circular orbit. When the satellite reaches apogee of the transfer orbit, another ΔV must be added or the satellite will remain in the transfer ellipse. This ΔV is the difference between the apogee velocity in the transfer orbit and the circular orbit velocity in the final orbit. After ΔV_f has been applied, the satellite is in the final orbit, and the transfer has been completed.

Derivation of Velocity Requirements and Transfer Time

Using the vis-viva equation and referring to Fig. 5.11,

$$V_1^2 = \mu \left[\frac{2}{r_1} - \frac{2}{r_1 + r_f} \right] \qquad V_{c1}^2 = \frac{\mu}{r_1}$$

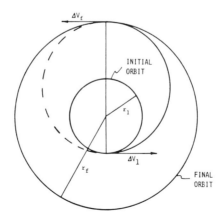

Fig. 5.11 Geometry of the Hohmann transfer.

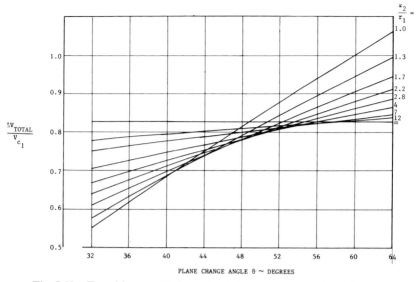

Fig. 5.19 Transition results for restricted three-impulse plane change.

Solving,

$$\rho_{OPT} = \frac{\sin\theta/2}{1 - 2\sin\theta/2} = \frac{1}{\cos\theta/2 - 2}$$

Upon examination, this equation reveals that, for $0 \leq \theta \leq 38.94$ deg, use

$$\frac{r_2}{r_1} = 1$$

for 38.94 deg $\leq \theta \leq 60$ deg, use

$$\frac{r_2}{r_1} = \frac{\sin\theta/2}{1 - 2\sin\theta/2}$$

for 60 deg $\leq \theta \leq 180$ deg, use

$$\frac{r_2}{r_1} \to \infty$$

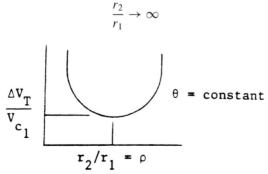

Fig. 5.20 Sketch of minimum $\Delta V_T/V_{c1}$ solution.

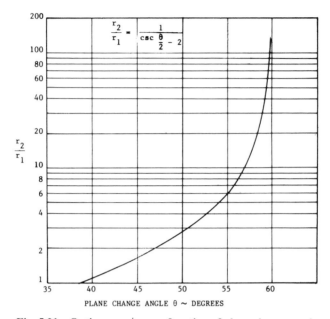

Fig. 5.21   Optimum r_2/r_1 as a function of plane change angle.

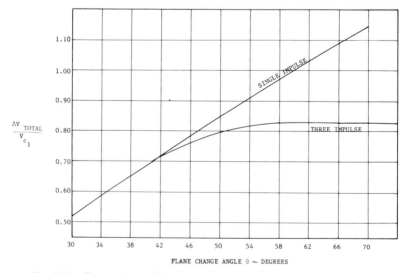

Fig. 5.22   Comparison of single- and restricted three-impulse maneuvers.

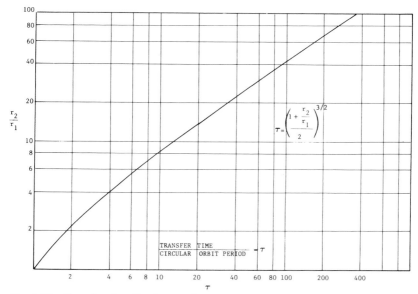

Fig. 5.23   Transfer time for the restricted three-impulse plane change maneuver.

Figure 5.21 presents a plot of optimum r_2/r_1 vs θ for 38.94 deg $\leq \theta \leq$ 60 deg. If these values are then substituted into the ΔV_T equation, they produce the minimum $\Delta V_T/V_{c1}$ as a function of θ. These results are presented in Fig. 5.22 and compared with the single-impulse results. The two curves are the same for $0 \leq \theta \leq 38.94$ deg. Figure 5.23 presents the transfer time divided by the initial orbit period as a function of r_2/r_1.

As an example, a plane change of 55 deg would prescribe an optimum value of $r_2/r_1 = 6$ from Fig. 5.21. From Fig. 5.22, minimum $\Delta V_T/V_{c1} = 0.820$. From Fig. 5.23, transfer time/circular orbit period = 6.6.

5.7   General Three-Impulse Plane Change Maneuver for Circular Orbits

This maneuver is like the restricted maneuver just described in that the initial and final circular orbits have the same radius but are rotated through an angle and the maneuver utilizes three impulses. The general case, however, makes a plane change at each of the three ΔV applications. The sum of the plane changes equals the total required rotation θ. These are special cases, $r_f = r_1$, in Refs. 5 and 6. However, Ref. 7 presents an analysis that minimizes $\Delta V_{\text{TOTAL}}/V_{c1}$ by determining optimum values of r_2/r_1, i.e., intermediate apogee radius, and of plane change distribution among the three ΔV applications. By symmetry, the plane change performed as part of the first and third ΔV applications is the same.

Results from Ref. 7 are presented in Fig. 5.24 and compared with previous results for the single-impulse and restricted three-impulse maneuvers. The general three-impulse maneuver provides the best results, i.e., lowest $\Delta V_{\text{TOTAL}}/V_{c1}$, for all plane change angles from 0 to 60.185 deg. For small angles, the ΔV results are only slightly smaller than the ΔV results for a single-impulse maneuver. At $\theta = 60.185$

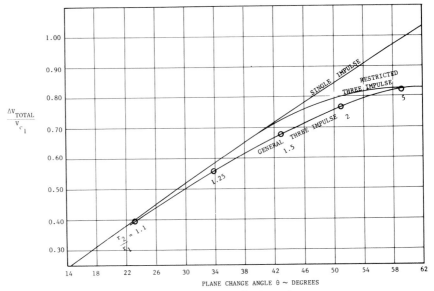

Fig. 5.24 Comparison of circular orbit plane change maneuvers.

deg, the general results merge with the results for a transfer to infinity and return, where $\Delta V_{\text{TOTAL}}/V_{c1} = 2(\sqrt{2} - 1)$. The optimum intermediate apogee radius ratio is noted for the circled points on the general three-impulse curve of Fig. 5.24. These values are generally lower than the corresponding values in Fig. 5.21 for the restricted three-impulse maneuver. This ratio is plotted on Fig. 5.25 as a function of θ. The interesting feature of this curve is that it remains very close to a value of 1 for relatively large values of θ until it increases dramatically as it approaches the value of $\theta = 60.185$ deg.

Figure 5.25 also presents a curve of the first plane change angle α_1 (or the third plane change angle α_3 since $\alpha_3 = \alpha_1$) as a function of θ, which equals $\alpha_1 + \alpha_2 + \alpha_3$. The value of α_1 reaches a maximum of about 4.85 deg at about $\theta = 49$ deg. For larger values of θ, α_1 then decreases, reaching 1.698 deg at $\theta = 60.185$ deg.

5.8 Hohmann Transfer with Split-Plane Change

An important practical circular orbit transfer is one that requires both a plane change and a radius change. The optimal two-impulse transfer to satisfy these requirements is the Hohmann transfer with plane change as shown in Fig. 5.26. The first ΔV not only produces a transfer ellipse whose apogee radius equals the final orbit radius, it also rotates the orbit plane through some angle α_1. At apogee, the second ΔV simultaneously circularizes the orbit and rotates the orbit plane through an angle α_2, where $\alpha_2 = \theta - \alpha_1$.

Figure 5.27 illustrates the velocity vector geometry for Hohmann transfer with plane change θ. Rider in Ref. 5 describes this as the "Mod-2 Hohmann" transfer. Baker in Ref. 6 describes this as the Hohmann transfer with plane change. Sketch 1

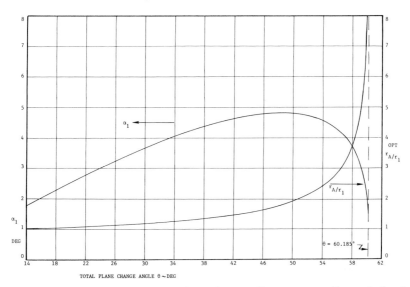

Fig. 5.25   Plane change angle and optimum intermediate apogee radius ratio for the general three-impulse plane change maneuver.

depicts the velocity vector triangle addition of ΔV_1.

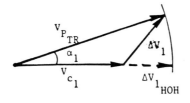

$$\Delta V_1^2 = V_{P_{TR}}^2 + V_{c1}^2 - 2V_{P_{TR}}V_{c1}\cos\alpha_1$$

$$V_{P_{TR}} = V_{c1} + \Delta V_{1_{HOH}}$$

$$V_{A_{TR}} = V_{c2} - \Delta V_{2_{HOH}}$$

Sketch 2 depicts the velocity vector triangle addition of ΔV_2.

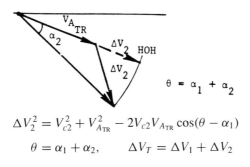

$$\Delta V_2^2 = V_{c2}^2 + V_{A_{TR}}^2 - 2V_{c2}V_{A_{TR}}\cos(\theta - \alpha_1)$$

$$\theta = \alpha_1 + \alpha_2, \qquad \Delta V_T = \Delta V_1 + \Delta V_2$$

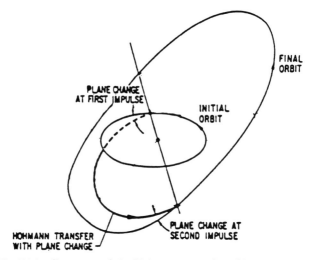

Fig. 5.26   Geometry of the Hohmann transfer with plane change.

To minimize ΔV_T, set $(\partial \Delta V_T / \partial \alpha_1) = 0$

$$\frac{\partial \Delta V_T}{\partial \alpha_1} = \frac{V_{P_{TR}} V_{c1} \sin \alpha_1}{\Delta V_1} - \frac{V_{c2} V_{A_{TR}} \sin(\theta - \alpha_1)}{\Delta V_2} = 0$$

Expanding

$$\frac{V_{P_{TR}} V_{c1} \sin \alpha_1}{\sqrt{V_{P_{TR}}^2 + V_{c1}^2 - 2V_{P_{TR}} V_{c1} \cos \alpha_1}}$$

$$= \frac{V_{c2} V_{A_{TR}} \sin(\theta - \alpha_1)}{\sqrt{V_{c2}^2 + V_{A_{TR}}^2 - 2V_{c2} V_{A_{TR}} \cos(\theta - \alpha_1)}}$$

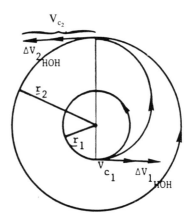

Fig. 5.27   Optimal plane split for Hohmann transfer with plane change.

Solve iteratively for $\alpha_{1_{OPT}}$. Then $\alpha_{2_{OPT}} = \theta - \alpha_{1_{OPT}}$. Substitute into the ΔV equations to get $\Delta V_{1_{OPT}}$ and $\Delta V_{2_{OPT}}$. Then,

$$\Delta V_{T_{MIN}} = \Delta V_{1_{OPT}} + \Delta V_{2_{OPT}}$$

5.9   Bi-elliptic Transfer with Split-Plane Change

This transfer is like the bi-elliptic transfer previously described except that plane changes are performed at each of the three ΔV applications. The geometry of the maneuver is presented in Fig. 5.28. Reference 5 describes this as the "Mod-2 bi-elliptic" transfer. Reference 6 describes this as the bi-elliptic transfer with plane change. Reference 7 presents an analysis that optimizes the plane change distribution among the three ΔVs and optimizes the intermediate apogee radius to minimize the total ΔV for a specified ratio of final to initial orbit radii and for a specified total plane change angle θ.

Results from Ref. 7 are presented in Fig. 5.29. In the space of final orbit radius/initial orbit radius and θ, Fig. 5.29 defines the best transfer modes, i.e., Hohmann transfer with plane change, bi-elliptic transfer with plane change, and parabolic transfer to infinity and return.

5.10   Transfer Between Coplanar Elliptic Orbits

Figure 5.30 illustrates two coplanar elliptic orbits, initial and final, and a transfer orbit between them. The transfer begins at a point of departure on the initial orbit. At this point, a velocity increment ΔV_1 is applied in some direction and added vectorially to the orbital velocity at that point. Thus, the satellite achieves a transfer orbit in which it coasts until it reaches an arrival point on the final orbit. At this point, a ΔV_2 that is the vector difference between the velocity in the final orbit and the velocity in the transfer orbit must be applied in order to complete the transfer.

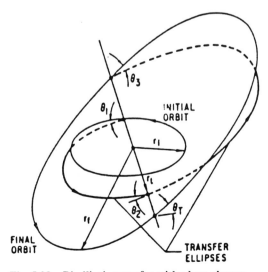

Fig. 5.28   Bi-elliptic transfer with plane change.

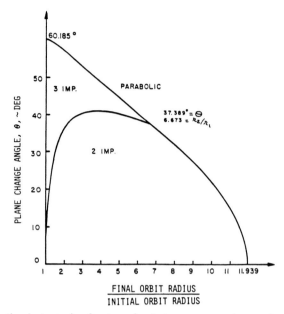

Fig. 5.29   Optimal strategies for transfer between noncoplanar circular orbits of radius r_1 and r_2 with optimal plane change distributions to minimize the total ΔV (from Ref. 7).

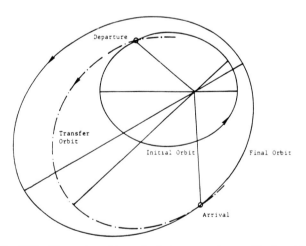

Fig. 5.30   Geometry of transfer between coplanar elliptic orbits.

Lawden, in Ref. 1, derives the equations for the optimal, i.e., minimum total ΔV, two-impulse transfer between coplanar elliptic orbits. These equations cannot be analytically solved in general. However, Lawden solved the special case of transfer between identical elliptic orbits that differ only in their orientation, as described in Sec. 5.3.

Other investigators have tried to solve this transfer problem by specializing the transfer to be a 180-deg transfer or to be cotangential. The 180-deg transfer restricts the departure and arrival points to be separated by 180 deg in central angle. The cotangential transfer requires the impulses to be tangent to the orbits at the points of application. These restrictions tend to simplify the problem of finding a minimum total ΔV solution. However, these solutions may not be as good as the general optimal solution and also are not easy to determine.

Bender, in Ref. 8, describes the theory and equations for the 180-deg transfer and the cotangential transfer. He also formulates a "practically optimum" transfer, which is a 180-deg circumferential transfer. Circumferential means that the ΔVs are applied in the circumferential direction, i.e., perpendicular to the radius vector. This maneuver is simple and easy to evaluate and provides good results. Comparisons of results of several transfer methods are presented in Ref. 8.

A very comprehensive survey of impulsive transfers is given in Ref. 9, which includes a listing of 316 references.

References

[1]Lawden, D. F., "Impulsive Transfer Between Elliptical Orbits," *Optimization Techniques,* edited by G. Leitmann, Academic, New York, 1962, Chap. 11.

[2]Hohmann, W., "Die Ereichbarkeit der Himmelskorper (The Attainability of Heavenly Bodies)," NASA, Technical Translations F-44, 1960.

[3]Hoelker, R. F., and Silber, R., "The Bi-Elliptical Transfer Between Coplanar Circular Orbits," *Proceedings of the 4th Symposium on Ballistic Missiles and Space Technology,* Vol. III, Pergamon, New York, 1961, pp. 164–175.

[4]Rider, L. A., "Characteristic Velocity for Changing the Inclination of a Circular Orbit to the Equator," *ARS Journal,* Vol. 29, Jan. 1959, pp. 48–49.

[5]Rider, L. A., "Characteristic Velocity Requirements for Impulsive Thrust Transfers Between Non-Coplanar Circular Orbits," *ARS Journal,* Vol. 31, March 1961, pp. 345–351.

[6]Baker, J. M., "Orbit Transfer and Rendezvous Maneuvers Between Inclined Circular Orbits," *Journal of Spacecrafts and Rockets,* Vol. 3, 1966, pp. 1216–1220.

[7]Hanson, J. H., "Optimal Maneuvers of Orbital Transfer Vehicles," Ph.D. Dissertation, Univ. of Michigan, Ann Arbor, MI, 1983.

[8]Bender, D. F., "Optimum Coplanar Two-Impulse Transfers Between Elliptic Orbits," *Aerospace Engineering,* Vol. 21, Oct. 1962, pp. 44–52.

[9]Gobetz, F. W., and Doll, J. R., "A Survey of Impulsive Transfers," *AIAA Journal,* Vol. 7, May 1969, pp. 801–834.

Problems

5.1. Given two circular orbits:

Initial	Final
$r_1 = 6660$ km ($h_1 = 289$ km)	$r_f = 133{,}200$ km
$i = 30$ deg	$i = 0$ (equatorial)

calculate the component and total ΔVs for the following transfer techniques from the initial orbit to the final orbit:

a) Plane change and then Hohmann transfer:

b) Hohmann transfer and then plane change:

c) Hohmann transfer with plane change at apogee in a vectorial combination (two impulses):

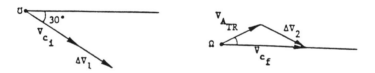

d) Bi-elliptic transfer with vectorial plane change at $r_t = 266,400$ km (three impulses):

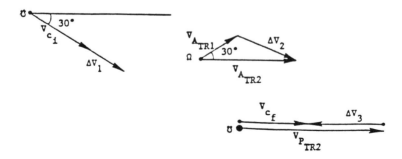

e) Hohmann transfer with optimally split-plane changes (two impulses)

to the nearest kilometer, Buck's minimum altitude during catch-up and Doc's maximum altitude during catch-up. How much time, in minutes, will Buck have to subdue Bart before Doc reaches them? Assume Venus to be spherical and atmosphereless. Assume the radius of Venus to be 6100 km and the circular orbit velocity at the surface of Venus to be 7210 m/s.

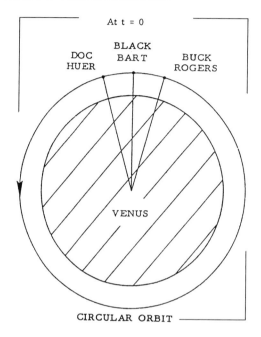

Selected Solutions

5.1. a) $\Delta V_1 = 4002$ m/s
$\Delta V_2 = 2939$ m/s
$\Delta V_3 = 1195$ m/s
$\Delta V_T = 8136$ m/s
b) $\Delta V_1 = 2939$ m/s
$\Delta V_2 = 1195$ m/s
$\Delta V_3 = 895$ m/s
$\Delta V_T = 5029$ m/s
c) $\Delta V_1 = 2939$ m/s
$\Delta V_2 = 1295$ m/s
$\Delta V_T = 4234$ m/s
d) $\Delta V_1 = 3068$ m/s
$\Delta V_2 = 776$ m/s
$\Delta V_3 = 267$ m/s
$\Delta V_T = 4112$ m/s

e) $\alpha_{1/\text{OPT}} = 0.7$ deg $\Delta V_1 = 2941$ m/s
   $\alpha_{2/\text{OPT}} = 29.3$ deg $\Delta V_2 = 1291$ m/s
   $\Delta V_T = 4232$ m/s

5.2. $\Delta V_{ABCD} = 1036$ m/s
   $\Delta V_{AD} = 991$ m/s
   $\Delta V_{ABE} = 1036$ m/s

5.5. $a = r$
   $\cos v = -e$
   $e = \sin \gamma$

5.6. $\Delta V_1 = 11,180$ m/s
   $V_2 = 7905$ m/s
   $V_{c2} = 5590$ m/s
   $\Delta V_2 = 5590$ m/s
   $\gamma_2 = 45$ deg
   $\alpha = 90$ deg

5.9. same
   same
   $r_{2i} = r_{2f}$

5.10. $\Delta V_1 = 2073$ m/s
   $\Delta V_2 = 2015$ m/s

5.11. Altitude = 64.2 km
   $\Delta V_{\text{BUCK}} = 37.6$ m/s retro
   $\Delta V_{\text{DOC}} = 36.5$ m/s forward
   Buck's minimum altitude = −63.5 km (underground)
   Doc's maximum altitude = 191.2 km
   time = 2.8 min if Buck's minimum altitude coincides with a deep canyon.

6
Complications to Impulsive Maneuvers

Having considered one-, two-, and three-impulse ΔV optimal transfers between orbits in Chapter 5, we now turn our attention to the complications of the real world. Are there cases in which four or more impulses would offer significant savings? Do we always want to use the ΔV optimal transfer? How bad is our assumption of impulsive ΔV application, and how does it affect the results? These are the questions that will be investigated in this chapter.

Specifically, the following topics will be considered: 1) N-impulse maneuver, 2) fixed-impulse transfers, 3) finite-duration burns, and 4) very low thrust transfers.

6.1 N-Impulse Maneuvers

In previous chapters, we have seen cases in which a two-impulse maneuver can accomplish the same transfer as a single-impulse but at a considerable ΔV saving (e.g., argument of perigee change). Similarly, cases were studied in which a three-impulse maneuver was best (e.g., bi-elliptic transfers). The obvious next question is: What about four or more impulses?

Edelbaum[1] has answered this question exactly in his paper "How Many Impulses?" The answer, he finds, is that only in rare situations are more than three impulses required to obtain the minimum ΔV transfer.

6.2 Fixed-Impulse Transfers

Introduction

To date, nearly all of the satellites that have been placed in low Earth orbit by the Space Shuttle have achieved their final mission orbit using solid-propellant rocket motors. Whether they are inertial-upper-stage (IUS) or payload-assist-module (PAM) engines, their operation is basically the same. Each solid rocket motor can be viewed simply as a container of solid propellant with a nozzle. An igniter begins the burning of the solid propellant, which continues until all propellant has been consumed. Whereas it is technically feasible to quench the burning of the solid propellant, the typical solid motors in use today are neither stoppable nor restartable.

Since most orbital transfers call for two burns, the typical upper-stage vehicle must consist of two solid motors as depicted in Fig. 6.1. Stage I provides the first burn (ΔV_1) and stage II provides the second burn (ΔV_2).

For a given payload weight and fixed (off-the-shelf) solid rocket motors, the ΔV available from the two motors are fixed by the rocket equation as derived in Chapter 1:

$$\frac{W_i}{W_f} = \exp(\Delta V / g_0 I_{sp}) \qquad (6.1)$$

117

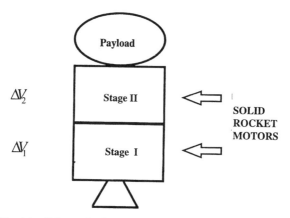

Fig. 6.1   Schematic diagram of a payload atop two solid rocket motors.

where

g_0 = the acceleration of gravity at the Earth's surface
I_{sp} = engine specific impulse
W_i = weight of rocket before burn (initial)
W_f = weight of rocket after burn (final)

Since the solid motors cannot be extinguished or restarted, the ΔV_1 and ΔV_2 supplied by the motors must exactly match the orbital-transfer ΔV requirements. The problem becomes one of finding a transfer between two given orbits that exactly matches the available ΔV_1 and ΔV_2. This problem is often referred to as the "velocity-matching technique" or "fixed-impulse transfers" (referring to the fact that the impulse available from each engine is fixed).

This is clearly a complication to the process of selecting an orbital transfer. In earlier chapters, the orbital transfer was sought that would minimize total ΔV expenditure. The inherent assumption was that the ΔV could be metered out in any size increments by the rocket engine.

Consider the plight of the designer who must assemble a pair of solid motors to form an upper-stage vehicle. In many cases, a range of different payload weights is to be carried, even if it is just to allow for a growth version of a single satellite. In this event, the designer must size the motors for the heaviest payload expected. Having done this, he must find a method to fly the smaller payloads with the same motors. Even the designer who has only a single satellite weight to contend with might be required to select his solid motors from the existing, off-the-shelf array. It is unlikely that he will find motors that will exactly match the ΔV of the optimal transfer. In any event, the designer's problem is the same: how to utilize solid motor stages where either one or both of the stages have excess energy above the optimal-transfer ΔV requirements. The problem is always excess energy because, if the sum of the ΔV provided by the solid motors is less than that of the optimal transfer, no solution is possible.

Several methods have been identified to accommodate this excess energy:
1) Offloading propellant.
2) Adding ballast.
3) Trajectory modification.

Before burn:

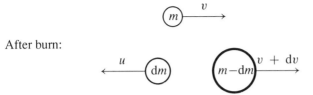

After burn:

It should be noted that the specific impulse of a rocket engine is a measure of the exhaust velocity of the particles according to the relation

$$u = g_0 I_{sp}$$

The momentum balance equation for this exchange was derived in Eqs. (1.5–1.8) as

$$m\frac{dv}{dt} = -u\frac{dm}{dt} + F$$

where F is the sum of all external forces acting on the masses.

The external forces acting on a launch vehicle are shown in Fig. 6.10. In this simple model, the thrust T and the drag D act along the axis of the vehicle while the weight W of the vehicle acts in the downward direction. The flight-path angle of the vehicle has been labeled Υ. Making the appropriate substitutions in the momentum equation yields

$$m\frac{dv}{dt} = -u\frac{dm}{dt} - D - mg\sin\Upsilon$$

Multiplying by dt/m, we obtain

$$dv = -u\frac{dm}{m} - \frac{D}{m}dt - g\sin\Upsilon\,dt$$

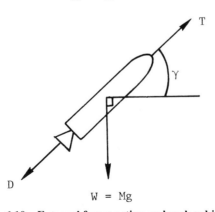

Fig. 6.10 External forces acting on launh vehicle.

Integrating from t_1 to t_2 (duration of the burn) yields*

$$\int_{v_1}^{v_2} dv \equiv \Delta V = -u \ln m \Big|_{M_1}^{M_2} - \int_{t_1}^{t_2} \frac{D}{m} dt - \int_{t_1}^{t_2} g \sin \Upsilon \, dt$$

or

$$\Delta V = g_0 I_{sp} \ln \frac{m_i}{m_f} - \int_{t_1}^{t_2} \frac{D}{m} dt - \int_{t_1}^{t_2} g \sin \Upsilon \, dt \qquad (6.5)$$

where

m_i = mass before burn (initial)
m_f = mass after burn (final)

The result is the familiar rocket equation [Eq. (1.10)], plus two complicating terms. In both cases, these terms detract from the rocket equation and reduce the available characteristic ΔV from the rocket. The first term is related to the drag force D and is not of interest here. In subsequent discussion, we shall assume $D = 0$.

The second term, the gravity-loss term, is the one we are interested in. We cannot easily integrate this term since both g (the gravitational acceleration at the rocket's location) and Υ (the flight-path angle) are unspecified functions of the time (which depend on the trajectory flown). We can, however, analyze this term with the intent of driving it toward zero.

Clearly, as t_1 approaches t_2, this gravity-loss term approaches zero. In the limit with $t_1 = t_2$, the burn is impulsive, and there are no gravity losses. The parameter g, we have no control over, except to note that burns at low altitude (high g) will be inherently more expensive than burns at high altitude (low g). We can control the flight-path angle Υ during the burn and, if it could be kept to zero, there would be no gravity losses. A finite-duration burn is shown in Fig. 6.11 with dashed lines. Typically, the burn surrounds the impulsive-transfer location (shown with solid lines), and the finite-burn direction is very close to the impulsive direction. Recall that, for Hohmann transfers, the flight-path angle is zero for both burns. If the impulsive-burn direction is to be maintained during the finite burn, the initial and final values of flight-path angle Υ cannot be zero. In fact, the initial values of Υ are approximately

$$-\Upsilon_1 \approx \Upsilon_2 \approx \phi_1 \approx \phi_2$$

so that the range of Υ is determined directly by the central angle through which the burn occurs. Near the midpoint of the burn arc, $\Upsilon \approx 0$ as desired but, at the ends, the $\sin \Upsilon$ contribution can be significant.

An alternative would be to vary the thrust direction during the burn so that, at any instant of time, $\Upsilon = 0$. By definition, the gravity-loss term would be zero in

---

*Note that

$$T = -u \frac{dm}{dt} = -g_0 I_{sp} \frac{dm}{dt} = -I_{sp} \frac{dW}{dt}$$

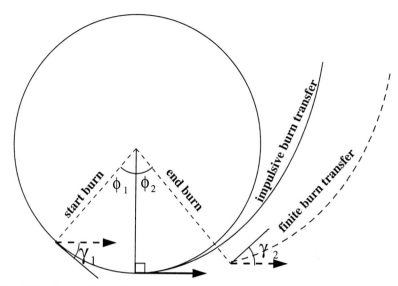

Fig. 6.11   Comparison of impulsive- and finite-burn transfers: solid lines = impulsive-burn characteristics; dashed lines = finite-burn characteristics.

Eq. (6.5). The fallacy in this approach is that the spacecraft would no longer follow the optimal-thrust direction so that the characteristic ΔV in Eq. (6.5) would be increased. One way or another, the penalty must be paid.

The factors that lead to high gravity losses are, then, long burn duration (low thrust/weight), high g (low altitude), and high Υ (large burn central angle). These gravity losses must be added to the impulsive ΔV to arrive at an effective ΔV to accomplish a maneuver. Figure 6.12 shows the total ΔV required to ascend from a low Earth orbit, LEO ($h = 278$ km, $i = 28.5$ deg) to a geosynchronous equatorial orbit, GEO ($h = 35786$ km, $i = 0$ deg) as a function of the initial thrust/weight ratio. For high thrust/weight ratios, the total ΔV approaches the impulsive ΔV (4237 m/s). For low thrust/weight ratios, the gravity loss is obviously significant (over 2200 m/s for $T/W_0 = 10^{-4}$).

Calculating Gravity Losses

For the purpose of determining gravity losses, the problem can be roughly divided into three categories.

High thrust $(T/W_0 \approx (0.5$ *to* $1.0)$. Here the thrust is the dominant force. Gravity losses can be estimated by using the methods of Robbins[4] or even neglected (impulsive assumption).

Low thrust $(T/W_0 \approx 10^{-2}$ *to* $10^{-1})$. Here the thrust and gravitational forces are both important, and assumptions or estimates are difficult. The best method of solution is to run an integrated trajectory computer program.

Very low thrust $(T/W_0 \approx 10^{-5})$. Here the thrust may be considered as a perturbation to the trajectory. An orbit transfer consists of many orbit revolutions

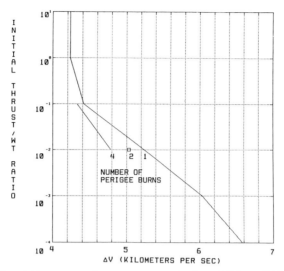

Fig. 6.12 ΔV requirements to transfer from LEO to GEO.

with continuous thrusting. A first-order analytic solution for this case will be described in the next section.

Reducing Gravity Losses

The obvious method for reducing gravity losses is to install an engine with a higher thrust. There is, however, another method. A single large (in terms of central angle traversed) burn can be broken up into a series of smaller burns, separated by one or more revolutions. Consider the transfer of Fig. 6.11. Rather than perform a single large burn over the central angle range $\phi_1 + \phi_2$, the burn could be split into two parts. In the first part, roughly half the total ΔV_1 would be applied, over roughly half the previous central angle range. After a full revolution, the vehicle would return to very nearly the same perigee location, where the remainder of the ΔV_1 would be applied. The net effect is that the central-angle travel (and the corresponding bounds on Υ) has been roughly halved. In the gravity-loss terms of Eq. (6.5), this means lower gravity losses. Breaking the burn into more and more pieces, each of which more closely resembles the impulsive case, leads to lower and lower gravity losses.

For the geosynchronous transfer of Fig. 6.12, some cases are plotted in which the large perigee burn has been broken into two or four pieces. The ΔV saving is substantial. The penalty paid for this is in the form of transfer time. The tradeoff between ΔV savings and transfer time for the case of $T/W_0 = 0.01$ is shown in Fig. 6.13 for the geosynchronous transfer. As the number of perigee burns increases, the gravity loss decreases, but the transfer time increases.

6.4   Very Low Thrust Transfers

If the thrust is small compared to the gravitational force $(T/W_0 \approx 10^{-5})$, then the resulting transfer orbit is a slow spiral outward under continuous thrust. That each of the many revolutions is nearly circular allows certain simplifying assumptions,

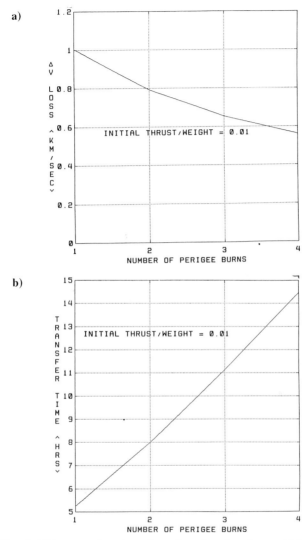

Fig. 6.13   a) ΔV loss as a function of number of perigee burns: b) transfer time as a function of number of perigee burns.

which have been used by Edelbaum[5] to develop a first-order analytic solution to the problem of very low thrust transfers between circular orbits.
   The given quantities are

V_0 = initial orbit circular velocity
V_1 = final orbit circular velocity
α = plane change angle between orbits
T = engine thrust

I_{sp} = engine specific impulse

W_1 = payload weight delivered to final orbit

The first step is to compute the ΔV required to make the transfer. This is given by Edelbaum as

$$\Delta V = \sqrt{V_0^2 - 2V_0V_1 \cos \frac{\pi}{2}\alpha + V_1^2}$$

Then, use the rocket equation to find the required weight in the initial orbit (W_0)

$$W_0 = W_1 \exp(\Delta V/g_0 I_{sp})$$

where g_0 is the gravitational acceleration at the Earth's surface. The propellant required for the transfer is simply

$$\Delta W = W_0 - W_1$$

The transfer time is, then,

$$\Delta t = \Delta W/\dot{W}$$

where

$$\dot{W} = \text{thrust}/I_{sp}$$

Using these equations, a first-order estimate of fuel and time requirements for a very low thrust upper-stage vehicle can be obtained. A more complete treatment of this problem is given in Chapter 14.

References

[1]Edelbaum, T. N., "How Many Impulses?," *Aeronautics and Astronautics,* Nov. 1967.

[2]Chu, S. T., Lang, T. J., and Winn, B. E., "A Velocity Matching Technique for Three Dimensional Orbit Transfer in Conceptual Mission Design," *Journal of the Astronautical Sciences,* Vol. 26, Oct.–Dec. 1978, pp. 343–368.

[3]Der, G. J., "Velocity Matching Technique Revisited," AAS/AIAA Astrodynamics Specialist Conference, Lake Tahoe, NV, Aug. 3–5, 1981.

[4]Robbins, H. M., "Analytical Study of the Impulsive Approximation," *AIAA Journal,* Vol. 4, Aug. 1966, pp. 1423–1477.

[5]Edelbaum, T. N., "Propulsion Requirements for Controllable Satellites," *ARS Journal,* Vol. 31, Aug. 1961, pp. 1079–1089.

Problems

6.1. For the fixed-impulse transfer example in the text, it is desired to increase the satellite weight from 907 to 990 kg. If this heavier satellite is placed atop the same upper-stage vehicle

a) Calculate the available ΔV_1 and ΔV_2.

b) Using these ΔV in a Hohmann-type transfer, what plane changes α_1 and α_2 must be performed?

c) What is the locus of Shuttle orbits (inclination and node) from which these fixed-impulse Hohmann-type, transfers can be achieved? Sketch in the locus onto Fig. 6.5.

d) Are non-Hohmann transfers available? Do they increase the allowable locus of Shuttle parking orbits?

e) Briefly describe the launch window available for a Shuttle parking orbit at $i = 55$ deg.

6.2. Instead of using two solid rocket motors to propel the satellite from the Shuttle parking orbit to the final orbit, let us now consider a very low thrust ion-propulsion engine:

$$\text{Thrust} = 4.45 \text{ N}$$

$$I_{sp} = 3000 \text{ s}$$

where $1 \text{ N} = 1 \text{ kg m/s}^2$ and the weight of 1 kg is $W = mg_0 = 1 \text{ kg} \times 9.8066$ m/s^2 = 9.8066 N
The payload weight to be placed on orbit now weighs 1361 kg and consists of the satellite, the ion engine, and empty propellant tanks.

a) For a Shuttle park orbit at $i = 55$ deg, what is the minimum plane change that must be performed by the ion upper stage?

b) How much ΔV is required from the ion engine?

c) How much gravity loss does this represent? (See example for optimal impulsive ΔV.)

d) How much propellant must the ion upper stage carry?

e) How long will the transfer take?

6.3. A payload of space experiments will be delivered to circular low Earth orbit (296-km altitude) by the Space Shuttle. Its final destination is a 741-km altitude circular orbit at as high an inclination as possible (preferably above $i = 72$ deg). Because of range safety constraints, the highest inclination available for the Shuttle orbit is $i = 57$ deg. The transfer from the Shuttle orbit to the destination orbit will be accomplished using two identical solid-propellant motors. These motors are not restartable, so that one motor will be consumed in leaving the Shuttle orbit and the second will be consumed in entering the destination orbit. The characteristics of the motors are as follows:

Payload = 1452 kg
Solid motor #2
    propellant wt = 1044 kg
    structure wt = 136 kg
    I_{sp} = 285 s

Solid motor #1
    propellant wt = 1044 kg
    structure wt = 136 kg
    I_{sp} = 285 s

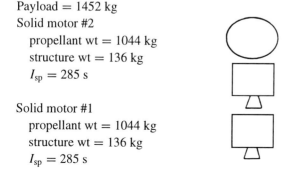

The required orbital transfer (a Hohmann-type transfer will be used) is summarized in the following sketch:

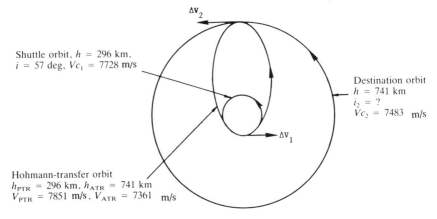

Shuttle orbit, $h = 296$ km, $i = 57$ deg, $Vc_1 = 7728$ m/s

Destination orbit $h = 741$ km $i_2 = ?$ $Vc_2 = 7483$ m/s

Hohmann-transfer orbit $h_{PTR} = 296$ km, $h_{ATR} = 741$ km $V_{PTR} = 7851$ m/s, $V_{ATR} = 7361$ m/s

For this orbital-transfer situation
   a) What $\Delta V (\Delta V_1$ and $\Delta V_2)$ are available from the two solid motors?
   b) How much plane change (α_1 and α_2) can be provided by each of the solid motors when flown on this transfer?
   c) What is the maximum value of inclination for the destination orbit using these motors? Have the experimenters met their goal of $i \geq 72$ deg?
   d) What is the transfer-orbit inclination for this case, and at what latitude must the two burns take place?

Selected Solutions

6.1.   a) $\Delta V_1 = 2048$ m/s
       $\Delta V_2 = 1779$ m/s
   b) $\alpha_1 \approx 0$
       $\alpha_2 = 19.97$ deg
   c) $\alpha_2 \approx 19.97$ deg

6.2.   a) $\Delta i = 0$
   b) $\Delta V = 3865$ m/s
   c) 395 m/s
   d) $m_p = 191$ kg
   e) $\Delta t = 14.7$ days

6.3.   a) $\Delta V_1 = 894$ m/s
       $\Delta V_2 = 1412$ m/s
   b) $\alpha_1 = 6.52$ deg
       $\alpha_2 = 10.87$ deg
   c) $i_{max} > 72$ deg
   d) $i_{tr} = 57 + 6.52 = 63.52$ deg

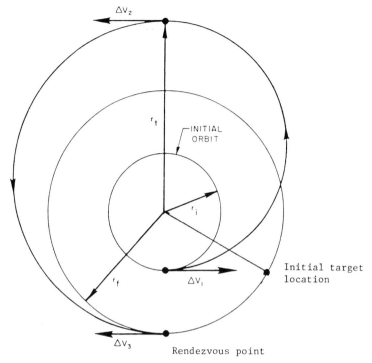

Fig. 7.4   Rendezvous via the bi-elliptic transfer.

and

$$t = t_H + \frac{P_f}{2}\left(3 - \frac{\Delta\theta}{\pi}\right) \qquad (7.8)$$

At $\Delta\theta = 0$,

$$t = t_H + \frac{3}{2}P_f$$

And at $\Delta\theta = 2\pi$,

$$t = t_H + \frac{P_f}{2}$$

Thus, because the bi-elliptic transfer occurs mostly beyond the outer circular orbit, it easily accommodates a $\Delta\theta$ that is slightly less than 2π. In this case, r_t will be only slightly larger than r_f.

In all cases, the value of r_t is determined by the value of $\Delta\theta$ because the time spent by the rendezvous satellite in the elliptic-orbit transfer legs is

$$t = \frac{\pi}{\sqrt{\mu}}\left[\left(\frac{r_i + r_t}{2}\right)^{3/2} + \left(\frac{r_t + r_f}{2}\right)^{3/2}\right] \qquad (7.9)$$

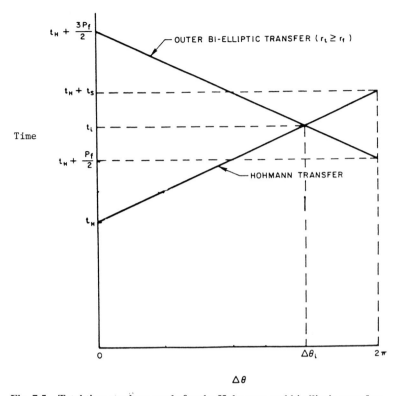

Fig. 7.5   Total time vs phase angle for the Hohmann and bi-elliptic transfers.

Figure 7.5 presents total time vs $\Delta\theta$. The Hohmann line is specified by Eq. (7.6). The bi-elliptic line is specified by Eq. (7.8). Note the difference in slope. The total time for the bi-elliptic transfer is a minimum at $\Delta\theta = 2\pi$. For smaller values of $\Delta\theta$, the total bi-elliptic time becomes longer. Therefore, there is no advantage to be gained by waiting because waiting reduces $\Delta\theta$, which increases the total bi-elliptic time.

Figure 7.5 depicts an intersection of the Hohmann and bi-elliptic lines. An intersection will exist only if $P_s \geq P_f/2$. Substituting for P_s,

$$\frac{P_i P_f}{P_f - P_i} \geq \frac{P_f}{2} \tag{7.10}$$

or

$$3P_i \geq P_f \tag{7.11}$$

Since period $P = 2\pi r^{3/2}/\sqrt{\mu}$, then,

$$r_f \leq 3^{2/3} r_i \tag{7.12}$$

Since $3^{2/3}$ is approximately 2.08, Eq. (7.12) determines that, for an initial orbit altitude of 100 n.mi. (185.2 km), the limit of usefulness for the bi-elliptic phasing technique is a final orbit altitude of 3927.7 n.mi. (7274.1 km). For final altitudes above this value, the Hohmann-transfer technique should be employed for all values of $\Delta\theta$.

When there is an intersection at $\Delta\theta_i$, the Hohmann technique would be used when $0 \le \Delta\theta < \Delta\theta_i$, and the bi-elliptic technique would be used when $\Delta\theta_i \le \Delta\theta \le 2\pi$. To find $\Delta\theta_i$, set the total Hohmann time to the total bi-elliptic time

$$t_H + \frac{\Delta\theta_i}{(2\pi/P_i) - (2\pi/P_f)} = t_H + \frac{P_f}{2}\left(3 - \frac{\Delta\theta_i}{\pi}\right) \tag{7.13}$$

Solving for $\Delta\theta_i$,

$$\Delta\theta_i = 3\pi\left(1 - \frac{P_i}{P_f}\right) \tag{7.14}$$

The corresponding total time is

$$t_i = t_H + \frac{3P_i}{2} \tag{7.15}$$

The break-even phasing angle $\Delta\theta_i$ is presented as a function of h_f for $h_i = 100$ n.mi. (185.2 km) in Fig. 7.6.

Semitangential transfer.   One more transfer technique to achieve coplanar rendezvous should be examined. Figure 7.7 illustrates the semitangential transfer. The rendezvous satellite achieves a transfer ellipse by applying a ΔV_1 that is larger than ΔV_1 for a Hohmann transfer. This transfer ellipse intersects the final circular orbit at two points, I_1 and I_2. Rendezvous can be accomplished at either point by the application of a second ΔV to circularize the orbit.

The rendezvous solution proceeds as follows:

1) Apply a specified ΔV_1 to V_{ci} in the direction of motion.
2) Knowing the perigee radius r_p and the perigee velocity V_p of the transfer ellipse, calculate the semimajor axis a, for the transfer ellipse from the energy equation

$$V_p^2 = \mu\left[\frac{2}{r_p} - \frac{1}{a}\right] \tag{7.16}$$

3) Calculate the eccentricity e of the transfer ellipse from

$$r_p = a(1 - e) \tag{7.17}$$

4) Calculate the true anomaly v_1 of the intersection point, I_1, from the orbit equation

$$r_f = \frac{a(1 - e^2)}{1 + e\cos v_1} \tag{7.18}$$

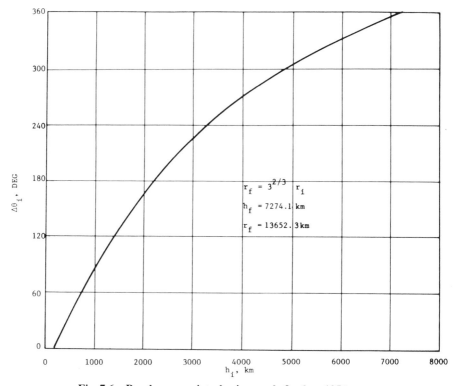

Fig. 7.6 Break-even point phasing angle for $h_i = 185$ km.

For I_2, $v_2 = 360$ deg $-v_1$,
5) Calculate the flight-path angle γ_1 from

$$\tan \gamma_1 = \frac{e \sin v_1}{1 + e \cos v_1} \qquad (7.19)$$

For I_2, $\gamma_2 = -\gamma_1$,
6) Calculate the eccentric anomaly E_1 from

$$\cos E_1 = \frac{e + \cos v_1}{1 + e \cos v_1} \qquad (7.20)$$

7) Calculate the period P_t of the transfer orbit from

$$P = \frac{2\pi a^{3/2}}{\sqrt{\mu}} \qquad (7.21)$$

8) Calculate the time t_1 from the application of ΔV_1 to the intersection point I_1 from Kepler's equation,

$$t_1 = \frac{P}{2\pi}(E_1 - e \sin E_1) \qquad (7.22)$$

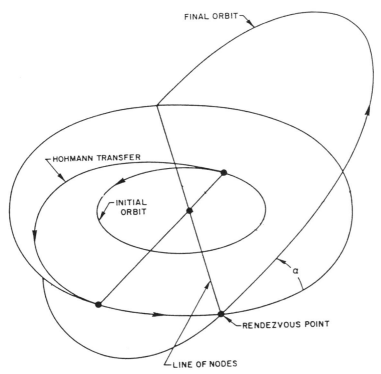

FINAL ORBIT

HOHMANN TRANSFER

INITIAL ORBIT

α

RENDEZVOUS POINT

LINE OF NODES

Fig. 7.15   Modified Hohmann-transfer maneuver.

and θ_f describes the position of the target satellite in the final orbit. Let $\theta_i - \theta_f = \theta_H + \Delta\theta$, as in the coplanar case. After a waiting time, $\Delta\theta$ becomes zero, and the Hohmann transfer is initiated. When the rendezvous satellite circularizes into the final orbit, both satellites are equidistant from the rendezvous point. When they simultaneously reach the rendezvous point, the rendezvous satellite performs a single-impulse plane change maneuver to rotate its orbit plane through the angle α, and rendezvous is accomplished. The time required for this technique is the sum of 1) the waiting time to achieve $\Delta\theta = 0$, 2) the Hohmann-transfer time, and 3) the time required for the rendezvous satellite to traverse the final orbit from the circularization point to the line of nodes.

Figure 7.16 describes the total velocity ΔV_T required for the three impulses as a function of the plane change angle α and the final circular orbit altitude h_f when the initial circular orbit altitude $h_i = 100$ n.mi. (185.2 km). When $\alpha = 0$ and $h_f = 300$ n.mi. (555.6 km), $\Delta V_T = 211$ m/s. This corresponds to the value shown on Fig. 7.10 for the coplanar Hohmann transfer. Note that ΔV_T increases very rapidly as α increases.

Bi-elliptic transfer with split plane changes.   This transfer technique was previously described in Chapter 5. For three-dimensional rendezvous, the bi-elliptic transfer is initiated when the rendezvous satellite reaches the line of nodes. As in the coplanar case, the value of $\Delta\theta$ determines the altitude h_t of the interme-diate transfer point. For the example of $h_i = 100$ n.mi. (185.2 km) and $h_f = 300$

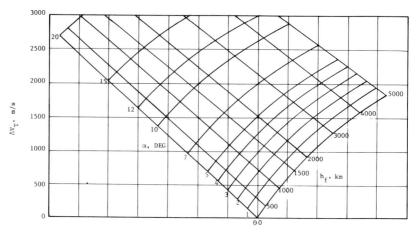

Fig. 7.16   Velocity required for a modified Hohmann transfer from a 185-km parking orbit.

n.mi. (555.6 km), Eqs. (7.8) and (7.9) determine h_t as a function of $\Delta\theta$. Then, given values for h_i, h_t, h_f, and the total plane change angle α_T, Ref. 8 describes the optimal plane change split, i.e., α_1, α_2, α_3, to minimize the total ΔV. Figure 7.17 presents these solutions for ΔV_T as a function of α_T and $\Delta\theta$ for $h_i = 100$ n.mi. (185.2 km) and $h_f = 300$ n.mi. (555.6 km). Because the ΔV_T for the coplanar bi-elliptic transfer is large compared to the Hohmann transfer (see Fig. 7.10), the advantage of the optimal split-plane change for the three-dimensional bi-elliptic transfer produces smaller values of ΔV_T than the modified Hohmann transfer only for large values of $\Delta\theta$.

In-Orbit Repositioning

Maneuvering technique.   If a satellite is to be repositioned in its circular orbit, this maneuver can be performed by applying an impulsive velocity along the

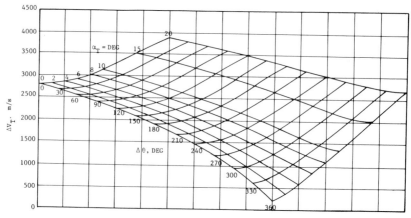

Fig. 7.17   Velocity increment ΔV_T necessary for a bi-elliptic transfer from a 185-km to a 556-km circular orbit.

velocity vector, either forward or retro. With a forward ΔV_1, the satellite will enter a larger phasing orbit. When the satellite returns to the point of ΔV application, it will be behind its original location in the circular orbit. The satellite can re-enter the circular orbit at this point by applying a retro ΔV equal in magnitude to the first ΔV. In a sense, a rendezvous with this point has been performed. Or the satellite can remain in the phasing orbit and re-enter the circular orbit on a future revolution. The satellite will drift farther behind with each additional revolution.

If the first ΔV is in the retro direction, the satellite will enter a smaller phasing orbit and will drift ahead of its original location in the circular orbit. The drift rate, either ahead or behind, is proportional to the magnitude of the ΔV.

Application to geosynchronous circular orbit.   A very common application of repositioning is the drifting of a satellite in a geosynchronous circular equatorial orbit from one longitude to another. The change in longitude is given by the equation

$$\Delta L = \dot{L} n P_{\mathrm{PH}} \tag{7.32}$$

where

ΔL = the change in longitude
$\dot{L}$ = the drift rate, positive eastward
n = the number of revolutions spent in the phasing orbit
P_{PH} = the period of the phasing orbit

The drift rate $\dot{L}$ is given by

$$\dot{L} = \omega_E \left(\frac{P_{\mathrm{PH}} - P_o}{P_{\mathrm{PH}}} \right) \tag{7.33}$$

where

ω_E = 360.985647 deg/day is the angular rate of axial rotation of the Earth
P_o = 1436.068 min = 0.9972696 days is the period of the
  geosynchronous orbit

Substituting Eq. (7.33) into Eq. (7.32),

$$\Delta L = \omega_E n (P_{\mathrm{PH}} - P_o) \tag{7.34}$$

The repositioning problem can now be addressed as follows. Given a desired longitudinal shift, say $\Delta L = +90$ deg, then, from Eq. (7.34),

$$n(P_{\mathrm{PH}} - P_o) = \frac{\Delta L}{\omega_E} = 0.249317 \text{ days} \tag{7.35}$$

Selecting a value of n allows the solution of $(P_{\mathrm{PH}} - P_o)$. Adding P_o solves for P_{PH}. Substitution of P_{PH}, n, and ΔL into Eq. (7.32) allows the solution of $\dot{L}$. Figure 7.18 is a graph of the ΔV required to start and stop a longitudinal drift rate

Table 7.1   Repositioning of geosynchronous satellite solutions for a
$\Delta L = +90$ deg

n rev	$P_{PH} - P_o$ days	P_{PH} days	$n P_{PH}$ days	$\dot{L}$ deg/day	ΔV m/s
6	0.04155	1.0388	6.233	14.44	82.0
12	0.02078	1.0181	12.217	7.37	41.8
24	0.01039	1.0077	24.184	3.72	21.3
96	0.00260	0.9999	95.987	0.94	5.49

in a geosynchronous orbit as a function of drift rate. Thus, a ΔV can be associated with a value of $\dot{L}$.

Table 7.1 presents a number of solutions to the $\Delta L = +90$ deg example. Four values of n were assumed. For each value n, the table presents values of P_{PH}, $n P_{PH}$ (the total elapsed time for repositioning), $\dot{L}$, and ΔV. If the repositioning is to be accomplished in 6 rev, then the drift rate is 14.44 deg/day, and the ΔV is 82.0 m/s. However, if the repositioning can be done slowly, i.e., in 96 rev, then the drift rate is only 0.94 deg/day, and the ΔV is only 5.49 m/s. This demonstrates the tradeoff between elapsed time and ΔV.

The curve on Fig. 7.18 was determined by assuming values of ΔV, calculating values of the phasing orbit semimajor axis from the energy equation, calculating

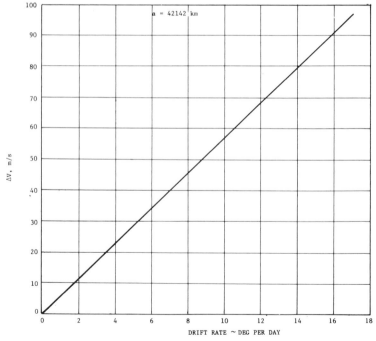

Fig. 7.18 ΔV required to start and stop a longitudinal drift rate in a geosynchronous circular orbit.

values of the phasing orbit period from the period equation, and calculating drift rates from Eq. (7.33). The equations and Fig. 7.18 work equally well for westward drifts.

7.2 Terminal Rendezvous

In the final phase of rendezvous before docking, the satellites are in close proximity, and the relative motion of the satellites is all-important. In this phase, it is common to describe the motion of one satellite with respect to the other. In the following subsections, the relative equations of motion will be derived. A solution to these equations will be obtained for the case in which one of the satellites is in a circular orbit.

Derivation of Relative Equations of Motion

Figure 7.19 presents the vector positions of the rendezvous and target satellites at some time with respect to the center of the Earth, r and r_T. The position of the rendezvous satellite with respect to the target satellite is ρ. An orthogonal coordinate frame is attached to the target satellite and moves with it. The y axis is radially outward. The z axis is out of the paper. The x axis completes a right-hand triad. The angular velocity, a vector of the target satellite, is given by ω.

The vector positions of the satellites yield

$$r = r_T + \rho \tag{7.36}$$

Differentiating this equation with respect to an inertial coordinate frame results in

$$\ddot{r} = \ddot{r}_T + \ddot{\rho} + 2(\omega \times \dot{\rho}) + \dot{\omega} \times \rho + \omega \times (\omega \times \rho) \tag{7.37}$$

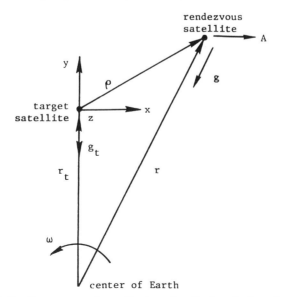

Fig. 7.19   Geometry and coordinate system for terminal rendezvous.

where, as in Eq. (1.2)

$\ddot{r}$ = the inertial acceleration of the rendezvous satellite
$\ddot{r}_T$ = the inertial acceleration of the target satellite
$\ddot{\rho}$ = the acceleration of the rendezvous satellite relative to
    the target satellite
$2(\omega \times \dot{\rho})$ = the Coriolis acceleration
$\dot{\omega} \times \rho$ = the Euler acceleration
$\omega \times (\omega \times \rho)$ = the centripetal acceleration

Now, let

$$\ddot{r} = g + A \tag{7.38}$$

where g is the gravitational acceleration and A the acceleration applied by external forces (thrust). Resolving Eqs. (7.37) and (7.38) into the x, y, and z components and solving for the relative accelerations produce

$$\ddot{x} = -g\frac{x}{r} + A_x + 2\omega\dot{y} + \dot{\omega}y + \omega^2 x$$

$$\ddot{y} = -g\left(\frac{y + r_T}{r}\right) + A_y + g_T - 2\omega\dot{x} - \dot{\omega}x - \omega^2 y \tag{7.39}$$

$$\ddot{z} = -g\frac{z}{r} + A_z$$

Assuming that the target-to-satellite distance is much smaller than the orbit radius of the target satellite or that

$$\rho^2 = x^2 + y^2 + z^2 \ll r_T^2 \tag{7.40}$$

the following approximate relations can be written

$$r = \left[x^2 + (y + r_T)^2 + z^2\right]^{1/2}$$

$$\approx r_T\left(1 + \frac{y}{r_T}\right)$$

$$g = \frac{g_T r_T^2}{r^2} \approx g_T\left(1 - \frac{2y}{r_T}\right) \tag{7.41}$$

$$-g\frac{x}{r} \approx -g_T\frac{x}{r_T}$$

$$-g\frac{z}{r} \approx -g_T\frac{x}{r_T}$$

$$-g\left(\frac{y + r_T}{r}\right) \approx -g_T\left(1 - \frac{2y}{r_T}\right)$$

Therefore, the linearized Eqs. (7.39) become

$$\ddot{x} = -g_T \frac{x}{r_T} + A_x + 2\omega\dot{y} + \dot{\omega}y + \omega^2 x$$

$$\ddot{y} = -2g_T \frac{y}{r_T} + A_y + 2\omega\dot{x} + \dot{\omega}x + \omega^2 y \qquad (7.42)$$

$$\ddot{z} = -g_T \frac{z}{r_T} + A_z$$

When the target is in a circular orbit, $\dot{\omega} = 0$ and $\omega = \sqrt{g_T/r_T}$, and Eqs. (7.42) become

$$\ddot{x} = A_x + 2\omega\dot{y}$$

$$\ddot{y} = A_y - 2\omega\dot{x} + 3\omega^2 y \qquad (7.43)$$

$$\ddot{z} = A_z - \omega^2 z$$

If there are no external accelerations (e.g., thrust), then,

$$A_x = A_y = A_z = 0$$

$$\ddot{x} - 2\omega\dot{y} = 0$$

$$\ddot{y} + 2\omega\dot{x} - 3\omega^2 y = 0 \qquad (7.44)$$

$$\ddot{z} + \omega^2 z = 0$$

Solution to the Relative Equations of Motion

The z equation is uncoupled from the x and y equations and can be solved separately. Assume a solution of the form

$$z = A \sin \omega t + B \cos \omega t \qquad (7.45)$$

Differentiating

$$\dot{z} = A\omega \cos \omega t - B\omega \sin \omega t$$

$$\ddot{z} = -A\omega^2 \sin \omega t - B\omega^2 \cos \omega t$$

When $t = 0$, $z = z_0$, and $\dot{z} = \dot{z}_0$, and so $z_0 = B$, and $\dot{z}_0 = A\omega$; therefore,

$$z = \frac{\dot{z}_0}{\omega} \sin \omega t + z_0 \cos \omega t \qquad (7.46)$$

$$\dot{z} = \dot{z}_0 \cos \omega t - z_0\omega \sin \omega t$$

Substitution into the $\ddot{z}$ equation verifies that these equations are a solution. In mechanics, they correspond to simple harmonic motion.

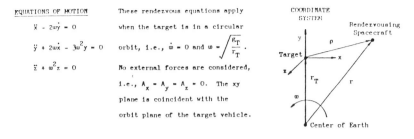

EQUATIONS OF MOTION

$\ddot{x} - 2\omega\dot{y} = 0$

$\ddot{y} + 2\omega\dot{x} - 3\omega^2 y = 0$

$\ddot{z} + \omega^2 z = 0$

These rendezvous equations apply when the target is in a circular orbit, i.e., $\dot{\omega} = 0$ and $\omega = \sqrt{\dfrac{g_T}{r_T}}$.

No external forces are considered, i.e., $A_x = A_y = A_z = 0$. The xy plane is coincident with the orbit plane of the target vehicle.

COORDINATE SYSTEM

x	1	$6(\omega t - \sin \omega t)$	0	$-3t + \dfrac{4}{\omega}\sin \omega t$	$\dfrac{2}{\omega}(1 - \cos \omega t)$	0	x_o
y	0	$4 - 3\cos \omega t$	0	$\dfrac{2}{\omega}(-1 + \cos \omega t)$	$\dfrac{1}{\omega}\sin \omega t$	0	y_o
z	0	0	$\cos \omega t$	0	0	$\dfrac{1}{\omega}\sin \omega t$	z_o
$\dot{x}$	0	$6\omega(1 - \cos \omega t)$	0	$-3 + 4\cos \omega t$	$2\sin \omega t$	0	$\dot{x}_o$
$\dot{y}$	0	$3\omega \sin \omega t$	0	$-2\sin \omega t$	$\cos \omega t$	0	$\dot{y}_o$
$\dot{z}$	0	0	$-\omega \sin \omega t$	0	0	$\cos \omega t$	$\dot{z}_o$

Fig. 7.20 Solution to the first-order circular-orbit rendezvous equations.

The x and y equations are coupled but can be solved to produce

$$x = x_0 + 2\frac{\dot{y}_0}{\omega}(1 - \cos \omega t) + \left(4\frac{\dot{x}_0}{\omega} - 6y_0\right)\sin \omega t + (6\omega y_0 - 3\dot{x}_0)t$$

$$y = 4y_0 - 2\frac{\dot{x}_0}{\omega} + \left(2\frac{\dot{x}_0}{\omega} - 3y_0\right)\cos \omega t + \frac{\dot{y}_0}{\omega}\sin \omega t \qquad (7.47)$$

$$\dot{x} = 2\dot{y}_0 \sin \omega t + (4\dot{x}_0 - 6\omega y_0)\cos \omega t + 6\omega y_0 - 3\dot{x}_0$$

$$\dot{y} = (3\omega y_0 - 2\dot{x}_0)\sin \omega t + \dot{y}_0 \cos \omega t$$

where x_0, $\dot{x}_0$, y_0, and $\dot{y}_0$ are position and velocity components at $t = 0$.

Figure 7.20 presents these solutions in matrix form. This is a compact, descriptive form. Given the initial position and velocity, the position and velocity at some future time can be determined from these equations.

Two-Impulse Rendezvous Maneuver

Given the initial position ρ_0 and velocity $\dot{\rho}_0$ for the rendezvous satellite with respect to the target satellite at the origin of the coordinate system and given the desire to rendezvous at a specified time τ, the problem is to find ΔV_1 at $t = 0$ and ΔV_2 at $t = \tau$ to accomplish rendezvous. Figure 7.21 presents a schematic of this two-impulse rendezvous maneuver.

The solution proceeds as follows.

If at time $t = 0$, the relative position x_0, y_0, z_0 is known (components of ρ_0), then the relative velocity components $\dot{x}_{0r}$, $\dot{y}_{0r}$, $\dot{z}_{0r}$ necessary to rendezvous at time

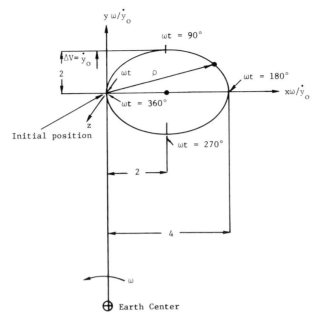

Fig. 7.26   Impulsive flyaround maneuver in orbit plane.

major axis is equal in magnitude to twice its minor axis and is proportional to the magnitude of the applied impulse. The solution is of the form

$$X = \frac{x\omega}{\dot{y}_0} = 2(1 - \cos \omega t)$$

$$Y = \frac{y\omega}{\dot{y}_0} = \sin \omega t$$

(7.52)

where $\dot{y}_0$ is the radial velocity impulse and ω is the orbital angular velocity. A plot of Eq. (7.52) is shown in Fig. 7.26. The relative displacement with respect to the origin of the coordinate reference frame is

$$\rho = (x^2 + y^2)^{1/2}$$

$$= \frac{\dot{y}_0}{\omega}[4(1 - \cos \omega t)^2 + \sin^2 \omega t]^{1/2}$$

(7.53)

Another approach for close circumnavigation is linear relative translation, which results when the orbit-dependent terms in the rendezvous equations are negligible. The applicable equations are

$$\ddot{x} = a_x, \qquad \ddot{y} = a_y$$

(7.54)

A circular flyaround at a constant radius can also be performed, which requires a continuous application of thrust to counteract the centrifugal acceleration.

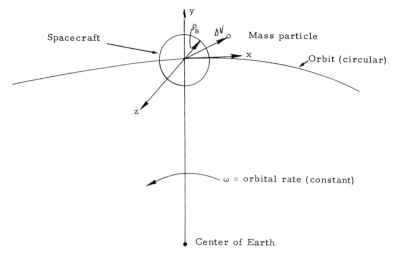

Fig. 7.27   Coordinate system.

Ejected Particle Trajectories

Collision with an ejected particle.   The possibility of a spacecraft collision with a particle ejected from a spacecraft is increased greatly if the particle is ejected radially or in a cross-track (out-of-plane) direction. The resulting motion of the particle, to a first-order approximation, is periodic in nature, implying that the particle will return to the ejecting body in an orbit period or a fraction thereof. The in-track (forward or backward) ejection, however, results in a secular increase of the particle distance from the ejecting body, which may or may not be large enough to avoid collision an orbit period later.

Consider the coordinate system attached to a satellite in a circular orbit, as shown in Fig. 7.27, where ρ_s is the radius of the satellite.

For a particle ejected with a relative velocity $\dot{x}_0$, $\dot{y}_0$, or $\dot{z}_0$ along the x, y, or z axes of the reference frame in Fig. 7.27, the trajectory equations from Fig. 7.20 are of the form

$$x = \left(-3t + \frac{4}{\omega} \sin \omega t \right) \dot{x}_0 + \frac{2}{\omega}(1 - \cos \omega t) \dot{y}_0$$

$$y = \frac{2}{\omega}(-1 + \cos \omega t)\dot{x}_0 + \frac{\dot{y}_0}{\omega} \sin \omega t \tag{7.55}$$

$$z = \frac{\dot{z}_0}{\omega} \sin \omega t$$

Solution of Eq. (7.55) leads to the following conclusions:

1) For radial ejection:
   a) Separation is periodic in time.
   b) Maximum separation occurs a half-orbit after ejection.
   c) Separation is reduced to zero upon completion of one orbit.
2) For tangential ejection:
   a) Separation is always finite and variable with time.

    b) In-track separation one orbit after ejection is maximum.

    c) Separation increases with succeeding orbits.

  3) For out-of-plane ejection:

    a) Out-of-plane ejection periodic in time (change in orbit inclination).

    b) Maximum separation occurs $\frac{1}{4}$ and $\frac{3}{4}$ of an orbit period after ejection.

    c) Separation reduced to zero every half-orbit.

For a particle ejected in an arbitrary direction from the satellite, the probability of recontact with the satellite would depend on the magnitudes of the tangential component of velocity $\dot{x}_0$. Thus, for example, one period later, i.e., when $\omega t = 2\pi$, the position of the mass (relative frame x, y, z) is given by the equation

$$x = -\frac{6\pi \dot{x}_0}{\omega} \qquad (7.56)$$

This result shows that the mass will be leading (negative x) the spacecraft for a backward ejection at the initial time $t = 0$ and lagging for a forward ejection at $t = 0$.

A sphere of radius $\rho_s = x$ can thus be defined as centered at the coordinate frame (spacecraft origin) that will not be entered by the ejected mass one orbit period later if $|\dot{x}_0| \geq \omega x/6\pi$. Consider now a given ejection velocity ΔV. The x component of ΔV can be defined as

$$|\dot{x}_0| = \Delta V \cos \beta \qquad (7.57)$$

where ΔV is the magnitude of the velocity vector, and β is a half-cone angle measured from the x axis, as shown in Fig. 7.28.

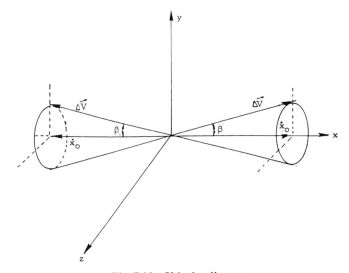

Fig. 7.28   Velocity diagram.

The $|\dot{x}_0| \geq \omega x/6\pi$ condition will be satisfied if and only if ΔV falls within the cone described by β (either along the positive or negative x axis), and the probability of this occurring can be expressed as

$$P\left(|\dot{x}_0| \geq \frac{\omega x}{6\pi}\right) = \frac{2A_z}{A_s} \tag{7.58}$$

where

A_z = an effective area of a spherical zone defined by the cone β
$\quad = 2\pi(\Delta V)^2(1 - \cos\beta)$
A_s = an effective spherical area
$\quad = 4\pi(\Delta V)^2$

assuming an equal probability of ΔV occurring along any direction. Thus, the probability that a mass initially ejected with a velocity ΔV in an arbitrary direction will be outside a sphere of radius ρ_s one orbital period following the ejection is

$$P = 1 - \cos\beta$$
$$= 1 - \frac{|\dot{x}_0|}{\Delta V}$$
$$= 1 - \frac{\omega x}{6\pi\Delta V} \tag{7.59}$$
$$= 1 - \frac{\omega\rho_s}{6\pi\Delta V}$$

The probability that the ejected mass will be within the sphere of radius ρ_s is then

$$P_{\rho_s} = 1 - P$$
$$= \frac{\omega\rho_s}{6\pi\Delta V} \tag{7.60}$$

Thus, for example, for a random ejection from a spacecraft with $\Delta V = 10$ m/s and $\rho_s = 100$ m, the probability of recontact (collision) in a 500-km circular orbit within a 100-meter radius is about 5.8×10^{-4}.

Debris cloud outline. The linearized rendezvous equations presented in Fig. 7.20 can be used to determine the outline of a debris cloud resulting from a breakup of a satellite in orbit. If, for example, it is assumed that the satellite breaks up isotopically; i.e., the individual particles receive a uniform velocity impulse ΔV in all directions, then the position of the particles can be computed as a function of time in an Earth-following coordinate frame attached to the center of mass of the exploding satellite to obtain the outline of the resulting cloud. This can be performed as follows:

Consider an explosion or a collision event in a circular orbit such as the one illustrated in Fig. 7.29. An orbiting orthogonal reference frame xyz is centered at the origin of the event at time $t = 0$ such that x is directed opposite to the

$$\ddot{y} = n^2 \left(\frac{a_{y_0}}{n^2} - d \right) \cosh nt$$

$$= (a_{y_0} - 3\omega^2 d) \cosh \sqrt{3}\omega t \tag{7.75}$$

The in-track acceleration a_x is then

$$a_x = -2\omega \dot{y}$$
$$= -\frac{2}{\sqrt{3}}(a_{y_0} - 3\omega^2 d) \sinh \sqrt{3}\omega t \tag{7.76}$$

For the case $t_1 \le t \le T$ when outward radial acceleration (thrusting) is zero, the specific solution of Eqs. (7.67) and (7.68) can be obtained from the conditions.

$$\dot{y}(T) = Ane^{nT} - Bne^{-nT} \rightarrow B = Ae^{2nT}$$
$$= 0$$
$$y(T) = Ae^{nT} + Ae^{2nT}e^{-nT}$$
$$= 2A(e^{nT}) = -\delta \rightarrow A = -\frac{\delta}{2}e^{-nT}$$

Thus, for $t_1 \le t \le T$,

$$y = -\frac{\delta}{2}\{\exp[n(t - T)] + \exp[-n(t - T)]\}$$
$$= -\delta \cosh n(t - T) \tag{7.77}$$
$$\dot{y} = -\delta n \sinh n(t - T)$$
$$= -\delta \sqrt{3}\omega \sinh \sqrt{3}\omega(t - T) \tag{7.78}$$
$$\ddot{y} = -\delta n^2 \cosh n(t - T)$$
$$= -3\omega^2 \delta \cosh \sqrt{3}\omega(t - T) \tag{7.79}$$

Consequently, for $t_1 \le t \le T$, $a_y = 0$ and

$$a_x = -2\omega \dot{y}$$
$$= (2n^2\delta/\sqrt{3}) \sinh n(t - T) \tag{7.80}$$

The total (combined) acceleration for $0 \le t \le t_1$ is

$$a_T = (a_x^2 + a_y^2)^{1/2}$$
$$= \left[\frac{4}{3}(dn^2 - a_{y_0})^2 \sinh^2 nt + a_{y_0}^2 \right]^{1/2} \tag{7.81}$$

The velocity impulse (ΔV) is given by

$$\Delta V = \int_0^T a_T \, dt \qquad (7.82)$$

The actual trajectory of the satellite, as obtained by integrating Eq. (7.63), will always contain an in-track $(-x$ direction) component. The magnitude of the in-track component can, however, be controlled by varying slightly the radial acceleration component a_{y_0}.

7.4 An Exact Analytical Solution for Two-Dimensional Relative Motion

Introduction

Interesting and worthwhile solutions for the relative motion of a probe, ejected into an elliptic orbit in the orbital plane of a space station that is in a circular orbit, are derived by Berreen and Crisp in Ref. 9. They have developed an exact analytical solution by coordinate transformation of the known orbital motions to rotating coordinates.

However, there are three restrictions on the solution of Berreen and Crisp that should be noted: 1) The probe is ejected from the space station at time $t = 0$ with relative velocity components x_0' and y_0' but the equations as derived do not permit an initial relative displacement such as $\rho_0 = (x_0^2 + y_0^2)^{1/2}$. Generalized equations that permit an initial relative displacement will be derived here. 2) The motion of the probe is restricted to the orbit plane of the space station and is, therefore, two-dimensional. 3) The space station is assumed to be in a circular orbit. As stated previously, restriction 1 will be relaxed in this section by the derivation of orbit element equations for the probe in terms of arbitrary initial relative velocity and displacement components for the probe with respect to the space station. The relaxation of restrictions 2 and 3 will be the subject of future studies.

Geometry and Coordinate Systems

Using the notation and description of Berreen and Crisp,[9] consider first the coordinate systems of Fig. 7.32. The space station–centered system is X, Y; the X_i, Y_i system is a geocentric inertial system: and the $X_e.Y_e$ system is a geocentric rotating system having its Y_e axis always passing through the space station. Coordinates R_p, θ_P and R_p, α are polar coordinates of the probe in the X_i, Y_i and X_e, Y_e systems, respectively.

Uppercase letters are used henceforth for real distances and velocities: and lowercase letters are used for ratios of distance and velocity, respectively, to the station orbital radius R_s and circular orbit velocity V_s. Thus,

$$x = \frac{X}{R_s}, \quad r_p = \frac{R_p}{R_s}, \quad v_p = \frac{V_p}{V_s} \qquad (7.83)$$

where

$$V_s = \sqrt{\frac{\mu}{R_s}} \qquad (7.84)$$

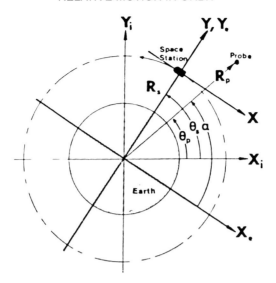

Fig. 7.32   The rectangular coordinate systems $(X_i, Y_i), (X_e, Y_e)$, and (X, Y) and the polar coordinates $(R_s, \theta_s), (R_p, \theta_p)$, and (R_p, α) (from Ref. 9).

and μ is the gravitational constant for the Earth. Subscripts s and p refer to station and probe, respectively.

The mean motion N_s of the station is

$$N_s = \frac{V_s}{R_s} \tag{7.85}$$

The angular coordinate θ_s of the station is then

$$\theta_s = N_s t \tag{7.86}$$

with initial conditions defined at $t = 0$.

Orbital Relations from Berreen and Crisp

In an inertial coordinate system, the probe moves in a Keplerian orbit described by the elements e_p, the eccentricity p_p, the semilatus rectum ratioed to R_s, and the apsidal orientation θ_p^*. Berreen and Crisp[9] describe these elements in terms of the initial relative velocity ratio components x_0' and y_0' by the equations

$$p_p = (1 - x_0')^2 \tag{7.87}$$

$$e_p^2 = 1 + p_p v_p^2 - \frac{2}{r_p} = 1 + (1 - x_0')^2 \left[y_0'^2 + (1 - x_0')^2 - 2 \right] \tag{7.88}$$

and

$$\theta_p^* = \cos^{-1}[(p_p - 1)/e_p] = -\sin^{-1}[(1 - x_0')y_0'/e_p] \tag{7.89}$$

with $-\pi < \theta_p^* \leq \pi$. The next subsection will generalize these equations to include initial relative displacement components x_0 and y_0, as well as x_0' and y_0'.

Derivation of Generalized Orbital Relations

Figure 7.33 depicts arbitrary, but still coplanar, initial conditions for the probe and space station.

From the geometry of Fig. 7.33,

$$r_p = \left[(1 + y_0)^2 + x_0^2\right]^{1/2} = \left(1 + 2y_0 + x_0^2 + y_0^2\right)^{1/2} \tag{7.90}$$

and

$$v_p = \left[(1 - x_0')^2 = y_0'^2\right]^{1/2} = \left(1 - 2x_0' + x_0'^2 + y_0'^2\right)^{1/2} \tag{7.91}$$

From conservation of angular momentum,

$$p_p = r_p^2 v_{p\text{horizontal}}^2 \tag{7.92}$$

where

$$v_{p\text{horizontal}} = 1 - x_0' \tag{7.93}$$

Substituting Eqs. (7.90) and (7.93) into Eq. (7.92).

$$p_p = \left(1 + 2y_0 + x_0^2 + y_0^2\right)(1 - x_0')^2 \tag{7.94}$$

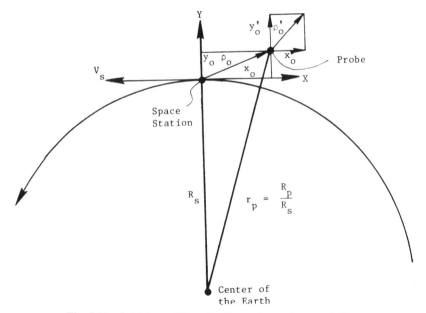

Fig. 7.33   Initial conditions for the probe and space station.

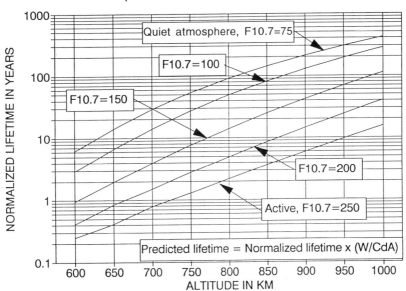

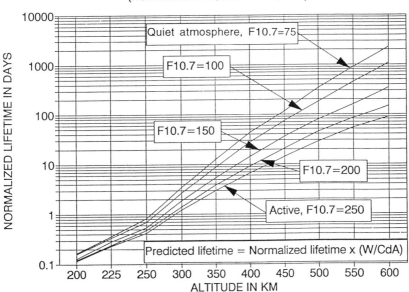

Fig. 8.5   Estimated orbit lifetime for average and active atmosphere.

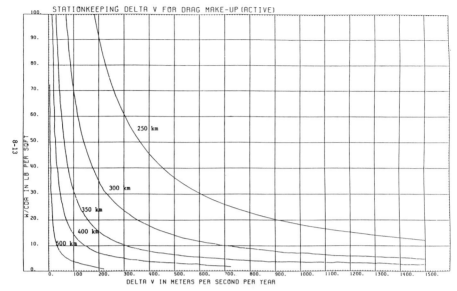

Fig. 8.6 **Drag makeup ΔV for an active Jacchia (1964) atmosphere ($F_{10.7} = 220, a_p = 15$) at various orbit altitudes and W/C_dA.**

responsible for the existence of some of the resonance phenomena discovered through the first telescope built by Galileo. These effects will be discussed in detail in Chapter 10.

References

[1]Roy, A. E., *Orbital Motion,* 3rd ed., Adam Hilger, Bristol, UK, 1988.

[2]Michielsen, H. J., and Webb, E. D., "Stationkeeping of Stationary Satellites Made Simple," *Proceedings of the First Western Space Conference,* 1970.

[3]Chao, C. C., "An Analytical Integration of the Averaged Equations of Variation Due to Sun-Moon Perturbations and Its Application," The Aerospace Corp., Tech. Rept. SD-TR-80-12, Oct. 1979.

9
Orbit Perturbations: Mathematical Foundations

In Chapter 8, the physical phenomena of orbit perturbations due to various sources have been discussed. This chapter provides an introduction to the mathematical foundations of those perturbations and the various methods of solution.

9.1 Equations of Motion

Two-Body Point Mass

Before going into equations of motion for orbit perturbations, it is important to review the two-body equations of motion in relative form. The equations of motion for a satellite moving under the attraction of a point mass planet without any other perturbations can be given in the planet-centered coordinates as

$$\frac{d^2 r}{dt^2} = -\mu \frac{r}{r^3} \tag{9.1}$$

where

r = position vector of the satellite
μ = gravitational constant
t = $time$

Equation (9.1) is a set of three simultaneous second-order nonlinear differential equations. There are six constants of integration. The solution of Eq. (9.1) can be either in terms of initial position and velocity: $x_0, y_0, z_0, \dot{x}_0, \dot{y}_0, \dot{z}_0$; or in terms of the six orbit elements: $a, e, i, \Omega, \omega, M$.

The closed-form conic solutions of the two-body equations of motion have been given in the earlier chapters, and they may be expressed in a general functional form as

$$r(t) = r(x_0, y_0, z_0, \dot{x}_0, \dot{y}_0, \dot{z}_0, t) \tag{9.2a}$$

or

$$r(t) = r(a, e, i, \Omega, \omega, M) \tag{9.2b}$$

Five of the six orbit elements $(a - \omega)$ in the preceding expression are constants, and M is the mean anomaly defined by

$$M = M_0 + n(t - t_0) \tag{9.3}$$

193

where

M_0 = mean anomaly at epoch, t_0

n = mean motion = $\sqrt{\dfrac{\mu}{a^3}}$

Figure 9.1 shows the orbit geometry of an orbiting satellite in the inertial Earth-centered equatorial coordinate system (ECI). It is important to know that, without perturbations, the orbit plane and perigee orientation stay fixed in the inertial space.

The orbit elements described earlier are called the classical orbit elements, and they are widely used in celestial mechanics. However, this set of elements becomes poorly defined and ill behaved when the eccentricity and/or the inclination become vanishingly small. To remedy this problem, a particular set of orbit elements was developed,[1] and they are defined as

$$
\begin{aligned}
a &= a \\
h &= e\sin(\omega + \Omega) \\
k &= e\cos(\omega + \Omega) \\
\lambda &= M + \omega + \Omega \\
p &= \tan(i/2)\sin\Omega \\
q &= \tan(i/2)\cos\Omega
\end{aligned}
\tag{9.4}
$$

Then, the solution may be expressed as

$$
\boldsymbol{r} = \boldsymbol{r}(a, h, k, \lambda, p, q)
\tag{9.5}
$$

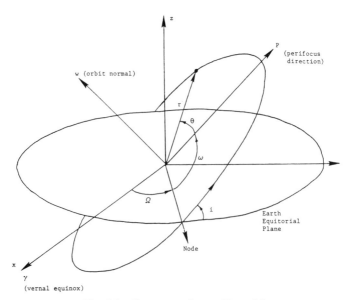

Fig. 9.1   Geometry of a satellite orbit.

where λ, the so-called mean longitude, is the only time-varying parameter, and

$$\lambda = \lambda_0 + n(t - t_0) \tag{9.6}$$

where

$$\lambda_0 = M_0 + \omega + \Omega = \text{mean longitude at epoch } (t_0)$$

Whether the two-body solutions are given in terms of initial position and velocity or orbit elements, one can always obtain the solutions in closed form. The new position and velocity can be computed at any given time.

In the real world, life is not that simple. Two-body solutions can give approximations for the orbit ephemeris for only a short time before the effect of perturbing accelerations becomes significant.

It is the purpose of this chapter to provide an overview of what the perturbations are and how these effects are computed by various methods.

Equations of Motion with Perturbations

To include the effects of the perturbations, the equations of motion can be written in a general form as

$$\frac{d^2 r}{dt^2} = -\mu \frac{r}{r^3} + a_p \tag{9.7}$$

where a_p is the resultant vector of all the perturbing accelerations. In the solar system, the magnitude of the a_p for all the satellite orbits is at least 10 times smaller than the central force or two-body accelerations, or $|a_p \ll |\mu r/r^3|$. a_p may consist of the following types of perturbing accelerations.

Gravitational:	Third-body (sun/moon) attractions
	The nonspherical Earth
Nongravitational:	Atmospheric drag
	Solar-radiation pressure
	Outgassing (fuel tank leaks on the spacecraft)
	Tidal friction effect

These perturbations can also be grouped as conservative and nonconservative. For conservative accelerations, a_p is an explicit function of position only, and there is no net energy transfer taking place. Therefore, the mean semimajor axis of the orbit is constant. For nonconservative accelerations, where a_p is an explicit function of both position and velocity, such as atmospheric drag, outgassing, and tidal friction effect, energy transfer occurs. Consequently, the mean semimajor axis of the orbit changes.

9.2   Methods of Solution

There are two general approaches to solve the equations of motion with perturbations. One approach is through a step-by-step numerical integration, which is often called "special perturbation." The other approach is through analytical expansion and integration of the equations of variations of orbit parameters. The latter approach is referred as "general perturbation."

Special Perturbation

In special perturbation, there are two basic methods: Cowell's method and Enckel's method. These two methods are explained in several books such as Refs. 2–4, and they are briefly introduced here.

Cowell's method. This method is a straightforward step-by-step integration of the two-body equations of motion with perturbations. The equations of motion may be given,

$$\ddot{r} + \frac{\mu}{r^3}r = a_p \tag{9.8}$$

which, for numerical integration, would be reduced to first-order differential equations

$$\begin{cases} \dot{r} = v \\ \dot{v} = -\dfrac{\mu}{r^3}r + a_p \end{cases} \tag{9.9}$$

where a_p is the vector sum of all the perturbing accelerations to be included in the integration. Cowell's method does not require that the magnitude of a_p should be small. Cowell's method can be further illustrated by the following example of integrating a second-order nonlinear differential equation of the form

$$\frac{d^2x}{dt^2} + 3x^2\frac{dx}{dt} - 5x = 12 \tag{9.10}$$

Equation (9.10) is first reduced to two first-order differential equations as

$$\begin{cases} \dfrac{dx}{dt} = y \\ \dfrac{dy}{dt} = -3x^2y + 5x + 12 \end{cases} \tag{9.11}$$

The step-by-step integration may take a simple recursive relation as

$$y(t_0 + \Delta t) = y(t_0) + \frac{dy}{dt}\bigg|_{t_0} \Delta t$$

$$x(t_0 + \Delta t) = x(t_0) + \frac{dx}{dt}\bigg|_{t_0} \Delta t \tag{9.12}$$

where $dx/dt|_{t_0}$ and $x(t_0)$ are given initial conditions and $y(t_0)$ and $dy/dt|_{t_0}$ can be computed from Eq. (9.11). The step size Δt should be so chosen that the round-off error and the truncation error are smaller than the error tolerance. Three types of integrators have been used in orbit computations: the Runge–Kutta method (fourth order and higher), the Adams–Moulton multistep predictor/corrector method, and the Gauss–Jackson (second sum) method. The Gauss-Jackson method integrates the second-order equations of motion directly without having to reduce them to first-order equations before integration. It was proved recently by Fox[5] that, for near-circular orbits, the Gauss–Jackson method with Herrick's[2] starting algorithm is the most efficient integrator for Cowell's method. These three methods have been programmed by the author in FORTRAN-77 to integrate satellite orbits with J_2 perturbations, and these programs are available in IBM/PC diskettes.

Encke's method.   One disadvantage of Cowell's method is slow computation. Even with the efficient Gauss–Jackson integrator, the computer time required to integrate an orbit to several hundred revolutions is still quite large. Encke's method cuts down the computer time considerably by integrating only the difference from a reference orbit whose ephemeris is known.

Encke's method uses a reference two-body orbit whose initial position and velocity equal that of the orbit with perturbations. Let the equation of motion of the reference orbit be

$$\ddot{\rho} = -\frac{\mu}{\rho^3}\rho \qquad (9.13)$$

and define the departure from this reference orbit as

$$\delta r = r - \rho \qquad (9.14)$$

where r must satisfy the equation of motion of the true orbit (Eq. 9.8). Then, by differencing Eq. (9.14) twice and with Eq. (9.13), one gets

$$\delta\ddot{r} = \mu\left(\frac{p}{\rho^3} - \frac{r}{r^3}\right) + a_p \qquad (9.15)$$

The difference δr is expected to be small and slowly varying as a result of the small perturbing accelerations a_p. Through some binomial expansions and approximations, Eq. (9.15) can be given as[4]

$$\delta\ddot{r} = \frac{\mu}{\rho^3}(fgr - \delta r) + a_p \qquad (9.16)$$

To integrate Eq. (9.16), the step size may be much larger than that of Cowell's method because of the slow-varying nature of δr. However, as the size of δr increases to a certain magnitude, the error in the approximation becomes too large, and a new reference orbit should be initialized. This procedure is called rectification, as shown in Fig. 9.2.

The advantages and disadvantages of these two methods in special perturbation may be summarized in Table 9.1.

General Perturbation

The integration of series, analytically, term by term, is the core of general perturbation theory. The integrand series are expansions based on the perturbing ac-

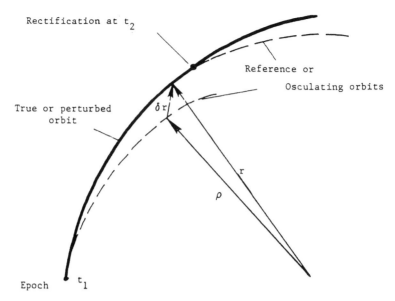

Fig. 9.2 Orbit geometry of Encke's method

celerations or the perturbing functions. These expansions are obtained through the methods of variation of parameters and variation of coordinates (Encke's method). Encke's method has been included in the special perturbation section; therefore, in this section, only the method of variation of parameters will be discussed.

Variation of parameters.   From the study of two-body motion, it is known that the coordinates and velocity components at any instant permit the determination of a unique set of six orbit elements. In the problem of two bodies, the elements do not change with time in the inertial reference coordinates. Hence, once the elements are determined at some epoch, the position and velocity of the satellite at any other time can be computed with the same number of significant figures as the basic data at epoch.

In the presence of perturbations, such as drag, third-body attraction, and Earth gravity harmonics, the Keplerian orbit elements are no longer constant. The concept of variation of parameters allows the orbit elements to vary in such a way that, at any instant, the coordinates and velocity components can be computed from a unique set of two-body elements as if there were no perturbations. The equations of the variations can be derived from the concept of perturbed variations. There are two basic approaches to obtain the variational equations in celestial mechanics. They are the force components approach and the perturbing function approach. The former is sometimes called the Gaussian method, and the latter is called the Lagrangian method.

The Force Components Approach

The force components approach directly relates the perturbing force components to the rate of the orbit elements. The general form of equations of variation can be

The basis of the development of the potential of a spheroid is in the integral

$$\Phi = k^2 \int \frac{dm}{s} \tag{9.32}$$

where dm is an element of the mass of the attracting body and s is the distance from this point-mass element to the attracted particle, a satellite S whose mass is assumed to be negligible (Fig. 9.3). Figure 9.3 shows two coordinate systems: 1) inertial coordinates (x, y, z) for the satellite motion, and 2) rotating coordinates (ξ, η, ζ) fixed to the attracting body. Through some lengthy mathematical integrations over the entire body, Eq. (9.32) becomes

$$\Phi = \frac{\mu}{r} \sum_{n=0}^{\infty} \sum_{q=0}^{n} \left(\frac{a_e}{r}\right)^n P_n^q(w)(C_{n,q} \cos q\lambda + S_{n,q} \sin q\lambda) \tag{9.33}$$

This infinite series is the potential function of a spheroid with coefficients $C_{n,q}$ and $S_{n,q}$ to be determined from observations, where

a_e = equatorial radius of the body
$P_n^q(w)$ = Legendre polynomials
w = $\sin\delta$, δ = declination of satellite
λ = longitude of the satellite in the body-fixed coordinates

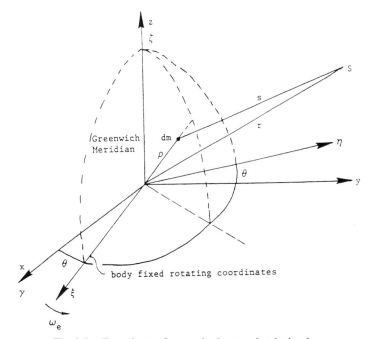

Fig. 9.3   Coordinates for gravity harmonics derivation.

The potential of a spheroid of revolution can be obtained from Eq. (9.33), with the center of mass at the center of the coordinate system as

$$\Phi = \frac{\mu}{r}\left[1 - \sum_{n=2}^{\infty}\left(\frac{a_e}{r}\right)^n J_n P_n(w)\right] \qquad (9.34)$$

where

$$J_n = -C_{n,0} = -C_n \qquad (9.35)$$

The mathematical definitions of $C_{n,q}$ and $S_{n,q}$ follow:
Zonal harmonics (spheroid of revolution about the spin axis):

$$C_{n,o} = C_n = -J_n = \frac{1}{ma_e^n}\int \rho^n P_n^o(W)\,dm \qquad (9.36)$$

The lowest-order zonal harmonics J_2 can be defined as

$$C_{2,0} = C_2 = -J_2 = \frac{1}{ma_e^2}\int \frac{\rho^2}{2}(3W^2 - 1)\,dm \qquad (9.37)$$

where

$\rho = (\xi^2 + \eta^2 + \zeta^2)$
m = mass of the spheroid
$W = \sin\phi'$
ϕ' = latitude of the mass element dm in the body-fixed coordinates

Tesseral and sectorial harmonics:

$$C_{n,q} = \frac{2}{ma_e^n}\frac{(n-q)!}{(n+q)!}\int \rho^n P_n^q(W)\cos q\lambda'\,dm \qquad (9.38)$$

where λ' is the body-fixed longitude of the mass element, and

$$C_{n,q} \rightarrow S_{n,q} \text{ for } \cos(\ ) \rightarrow \sin(\ ) \qquad (9.39)$$

The preceding definitions indicate that the values of the harmonics depend on the shape and mass distribution of the central body. For Earth, various sets of gravity harmonics have been determined in the past two decades. The most widely used set, which is probably the most accurate one, is the WGS 84 model. The values of the low-order harmonics (4×4) of that model are listed in Table 9.2 for reference. It is important to know that the value of J_2 is about 400 times larger than the next-largest value J_3. That is why, for most satellite orbits, a reasonably good accuracy can be maintained by simply including the J_2 effect.

9.4 More Definitions of Gravity Harmonics

Sometimes, it is more convenient to represent those harmonic coefficients in terms of J_{nm} and λ_{nm} instead of $C_{n,q}$ and $S_{n,q}$. Let us now review the mathematical

Table 9.2   Value's of low-order (fourth) harmonics of Earth gravitation potential (WGS 84 model)

Zonal harmonics
$J_2 = 1082.6300E\text{-}6$[a]
$J_3 = -2.5321531E\text{-}6$
$J_4 = -1.6109876E\text{-}6$
Tesseral harmonics

$C_{22} = 1.5747419E\text{-}6$	$S_{22} = -9.0237594E\text{-}7$
$C_{31} = 29.146736E\text{-}6$	$S_{31} = 2.7095717E\text{-}7$
$C_{32} = 3.0968373E\text{-}7$	$S_{32} = -2.1212017E\text{-}7$
$C_{33} = 1.0007897E\text{-}6$	$S_{33} = 1.9734562E\text{-}7$
$C_{42} = 7.7809618E\text{-}8$	$S_{42} = 1.4663946E\text{-}7$
$C_{44} = -3.9481643E\text{-}9$	$S_{44} = 6.540039E\text{-}9$

[a]$E\text{-}6 = \times 10^{-6}$

expressions of those definitions of the geopotential harmonics in a slightly different representation.

The external geopotential function at any point P specified by the spherical coordinates r, ϕ, λ defined in Fig. 9.4 can be expressed as

$$V = -\frac{\mu}{r}\left[1 - \sum_{n=2}^{\infty} J_n\left(\frac{R_e}{r}\right)^n P_n(\sin\phi)\right.$$

$$\left. + \sum_{n=2}^{\infty}\sum_{m=1}^{n} J_{nm}\left(\frac{R_e}{r}\right)^n P_{nm}(\sin\phi)\cos m(\lambda - \lambda_{nm})\right] \tag{9.40}$$

where

r = geocentric distance
ϕ = geocentric latitude
λ = geographic longitude
μ = GM_e = Newtonian constant times mass of Earth
R_e = mean equatorial radius of the Earth
P_{nm} = associated Legendre polynomial of degree n and order m
J_{nm} = harmonic coefficients
λ_{nm} = equilibrium longitude for J_{nm}
P_n = Legendre polynomial of degree n and order of zero
J_n = J_{n0}

where the relationships between J_{nm}/λ_{nm} and $C_{n,q}/S_{n,q}$ are

$$J_{nm}^2 = C_{n,q}^2 + S_{n,q}^2$$

$$\lambda_{nm} = \tan^{-1}(S_{n,q}/C_{n,q})/m, \qquad q = m$$

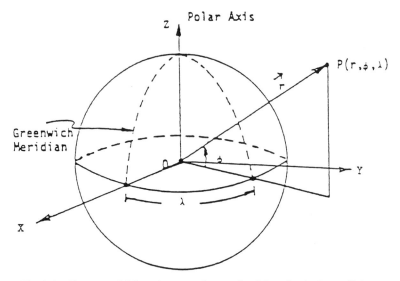

Fig. 9.4   Geopotential function at a given point P in spherical coordinates.

1) *Zonal Harmonics* ($m = 0$): For $m = 0$, the dependence of V on longitude vanishes, and the field is symmetric about the polar axis. For any $P_n(\sin \phi)$, there are n circles of latitude along which P_n is zero and, hence, $(n + 1)$ zones in which the function is alternately $(+)$ and $(-)$.

2) *Sectorial Harmonics* ($n = m$): The polynomial $P_{nm}(\sin \phi)$ are zero only at the poles ($\phi = \pm \pi/2$). On the other hand, the terms

$$\begin{pmatrix} \cos \\ \sin \end{pmatrix} n\lambda$$

are zero for $2n$ different values of λ; hence, the lines along which the function

$$\begin{pmatrix} \cos \\ \sin \end{pmatrix} n\lambda$$

$P_{nn}(\sin \phi)$ vanish are meridians of longitude, which divide the sphere into $2n$ "orange-slice" sectors alternately $(+)$ and $(-)$.

3) *Tesseral Harmonics* ($n \neq m$): For $n \neq m$, the functions

$$\begin{pmatrix} \cos \\ \sin \end{pmatrix} m\lambda$$

$P_{nm}(\sin \phi)$ are referred to as tesseral ("square") harmonics, for the sphere is divided up into a checkerboard array of domains alternately $(+)$ and $(-)$.

In general,

$$P_{nm}(x) = \frac{(1 - x^2)^{\frac{m}{2}}}{2^n n!} \cdot \frac{d^{(n+m)}(x^2 - 1)^n}{dx^{(n+m)}} \tag{9.41}$$

An illustration of the different zonal, tesseral, and sectoral harmonics (with positive and negative values) is shown in Fig. 9.5.

Table 9.3   Summary of special and general perturbation methods

	Special perturbation	General perturbation
Advantages	High precision General purpose Simplicity of formulation	Analytical Fast computation
Disadvantages	Slow computation	Orbit-type- dependent Time-consuming formulation
Applications	Orbit determination High-accuracy orbit Prediction Navigation	Mission design and analysis Long-term orbit Prediction Parametric study

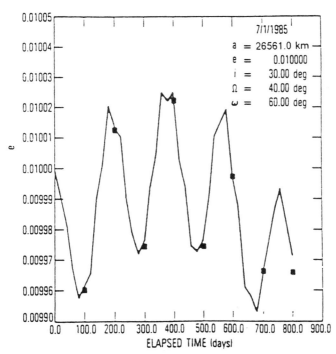

Fig. 9.7   Eccentricity time history for moderate inclination orbit (i = 30 deg).

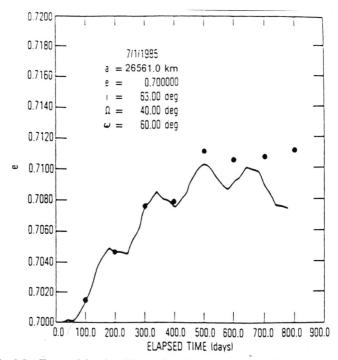

Fig. 9.8   Eccentricity time history for high-eccentricity orbit (i = 63 deg).

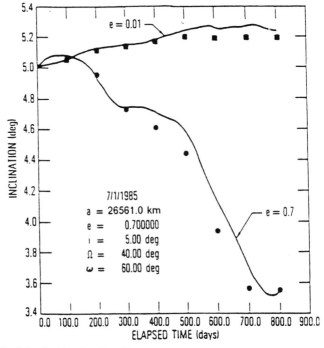

Fig. 9.9   Inclination time history for low-inclination orbit (i = 5 deg).

References

[1]Broucke, R. A., and Cefola, P. J., "On the Equinoctial Orbit Elements," *Celestial Mechanics,* Vol. 5, 1972.

[2]Herrick, S., *Astrodynamics,* Vol. II, Van Nostrand Reinhold, London, 1972.

[3]Brouwer, D., and Clemence, G., *Methods of Celestial Mechanics,* Academic, New York, 1961.

[4]Bate, R. R., Mueller, D. D., and White, J. E., *Fundamentals of Astrodynamics,* Dover, New York, 1971.

[5]Fox, K., "Numerical Integration of the Equations of Motion of Celestial Mechanics," *Celestial Mechanics,* Vol. 33, 1984.

[6]Chao, C. C., "A General Perturbation Method and Its Application to the Motions of the Four Massive Satellites of Jupiter," , Ph.D. Dissertation, UCLA, Los Angeles, CA, 1976.

[7]Chao, C. C., "An Analytical Integration of the Averaged Equations of Variation Due to Sun-Moon Perturbations and Its Application," , The Aerospace Corp., El Segundo, CA, Rept. SD-TR-80-12, Oct. 1979.

[8]Liu, J.J.F., and Alford, R. L., "Semianalytic Theory for a Close-Earth Artificial Satellite," *Journal of Guidance and Control,* Vol. 3, July–Aug. 1980.

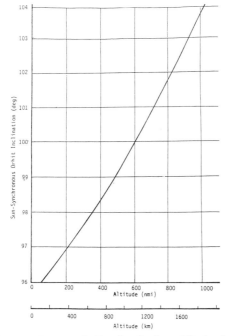

Fig. 10.5   Sun-synchronous inclination for low-altitude circular orbits.

A typical sun-synchronous orbit has a 98.7-deg inclination and 833 km mean orbit altitude. For low-altitude circular orbits, the sun-synchronous inclination and altitude are computed from Eq. (10.4) and plotted in Fig. 10.5. A plot for high altitude orbits is given in Chap. 11. The unique property of the sun-synchronous orbits is that the satellite's ground track has one local time on its ascending half and another local time (12 h away) on its descending half. The two local times remain the same for the entire mission. The sun-synchronous orbits are often referred to by the local time of the ascending node as 6 a.m. orbit, 10 a.m. orbit, etc. The local times at various satellite subpoints of the ground track can be computed by the following relation:

$$t = t_{node} - \{\tan^{-1}[\tan u \cos(\pi - i)]\} \div \omega_e \qquad (10.5)$$

where

t_{node} = local time of the ascending node crossing
u = argument of latitude of the satellite
i = inclination of the orbit
ω_e = Earth rotation rate

There are two kinds of perturbations, the drag and sun's attraction, that will affect the sun-synchronous property and gradually change the local time. The drag perturbs the orbit parameters p and n in Eq. (10.4), and the sun's attraction perturbs the orbit inclination i. The sun-synchronous property induces the deep resonance in the equation of the inclination variation due to solar perturbations.

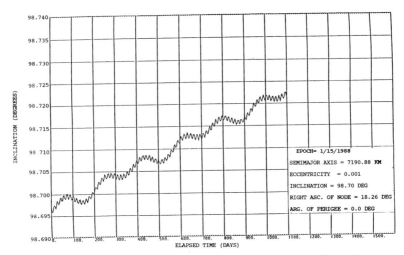

Fig. 10.6   Inclination variation without correction (5:30 orbit).

The particular term of the deep resonance that causes a drift in inclination may be given as[1]

$$\frac{\mathrm{d}i}{\mathrm{d}t} = -0.047 (\mathrm{deg/year}) \sin 2(\alpha_\odot - \Omega) \qquad (10.6)$$

with

$$\alpha_\odot = \text{right ascension of the sun}$$

and

$$\dot{\alpha}_\odot = \dot{\Omega} \ (\text{deep resonance})$$

The largest inclination drift is 0.047 deg/yr when the angle $2(\alpha_\odot - \Omega)$ is either 90 or 270 deg. Figure 10.6 shows a 3-yr inclination variation without correction for the nominal N-ROSS orbit with 5:30 a.m. local time. The corresponding value of $(\alpha_\odot - \Omega)$ is 277.5 deg, and the inclination rate as computed from the preceding equation is 0.012 deg/yr, which agrees with the mean slope of the inclination variation in Fig. 10.6. The inclination history shown in Fig. 10.6 was generated from high-precision numerical integration. For a further discussion of sun-synchronous orbits, see Chap. 11.

10.4 J_3 Effects and Frozen Orbits

The third harmonic J_3 is of the order of $10^{-3} J_2$ for the Earth, so that the amplitudes of the short-period perturbations are very small. However, in the equations of long-term variations for eccentricity, the eccentricity appears in the denominator of one term and thus will increase the magnitude of the term to nearly the magnitude of J_2 if the eccentricity is small enough. As a result, the eccentricity will have a long-term variation of the following form[1]:

$$\Delta e = -\frac{1}{2}\frac{J_3}{J_2}\left(\frac{R}{a}\right)\sin i \sin \omega \tag{10.7}$$

Equation (10.7) indicates that the J_3-induced variation in eccentricity will have a sinusoidal oscillation with a period of $2\pi/\dot{\omega}$ and an amplitude of $(J_3/J_2)(R/a)$, where R is the Earth equatorial radius. Similarly, there is a corresponding long-term variation in inclination with the following form[1]:

$$\Delta i = \frac{1}{2}\frac{J_3}{J_2}\left(\frac{R}{a}\right)\frac{e}{1-e^2}\cos i \sin \omega \tag{10.8}$$

It is seen that the induced inclination variation will become significant when the eccentricity is not too small and the inclination is not too large.

Through the coupling effect of J_2 and J_3, the concept of frozen orbit was introduced.[19] The frozen orbit can be achieved by properly selecting a particular combination of initial eccentricity and inclination for a given orbit period such that there will be no variation to the order of J_3 in eccentricity and argument of perigee. In other words, the orbit will be "frozen" in the inertial space, and the orbit altitude history will repeat every revolution. This property may be affected by the solar-radiation pressure and drag perturbations. Thus, in actual application, small orbit maintenance maneuvers will have to be performed periodically to offset those perturbations. For examples of frozen orbits, see Chap. 11.

10.5   Earth's Triaxiality Effects and East-West Stationkeeping

The Earth cross section along the equatorial plane is not a circle but more like an ellipse, as shown in Fig. 10.7. This ellipse-shaped cross section is represented by the so-called tesseral harmonics $(C_{22}, S_{22}, C_{32}, S_{32} \ldots)$. The primary tesseral harmonic is designated by J_{22}, which combines C_{22} and S_{22}. The longitude of symmetry of the J_{22} harmonic denoted by λ_{22} is determined from observations and has a typical value of -14.7 deg. The equilibrium points are divided into stable (75.3°E and 255.3°E) and unstable (14.7°W and 165.3°E) longitudes. The resonance effect of the 24-h geosynchronous orbits induces a very slow motion,

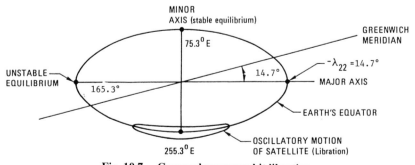

Fig. 10.7   Geosynchronous orbit libration.

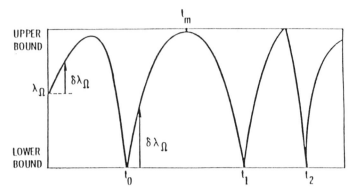

Fig. 10.8   Schematic drawing of a longitude stationkeeping method.

librating around the stable longitudes with a period of about 800 days and an amplitude of nearly 90 deg.

Methods of east-west stationkeeping were developed and discussed in Refs. 8, 9, 11, and 12. The concept is to apply a ΔV maneuver at t_0, where the longitude is about to exceed the tolerance boundary, as shown in Fig. 10.8. The magnitude of the ΔV is determined such that the resonance effect will bring the satellite back before it exceeds the other bound at t_m. The typical longitude tolerance is ± 1 deg, and the average time between two maneuvers is about 100 days. A much tighter tolerance of ± 0.1 deg is required by some missions. The stationkeeping ΔV per year is a function of longitude and can be found in Fig. 8.1.

10.6   Third-Body Perturbations and North/South Stationkeeping

Perturbations due to Earth's oblateness become less important with increasing distance from the Earth. On the other hand, the perturbations from the gravitational force of the moon and sun become more important at higher altitudes.

The combined effects of the solar and lunar gravitational attraction on high-altitude Earth orbits have been studied analytically by several investigators.[2-7] The combined lunisolar gravitational perturbations will induce long-term and/or secular variations in high-altitude orbit eccentricity, inclination, node, and argument of perigee.

The orbit inclination perturbations are caused primarily by the attractions of the sun and moon. For a geosynchronous orbit with initial zero inclination, the sun/moon perturbations increase the orbit inclination at a rate of about 1 deg/yr for the first 10 years, and then the inclination reaches 15 deg in about 17 yr.[10] After that, the inclination decreases to zero in another 27 yr. For most geosynchronous satellite missions, the orbit inclination is not controlled, primarily because of the costly ΔV expenditure in plane change maneuvers. Instead, the initial inclination and right ascension of the ascending node are properly chosen such that the inclination will first decrease to zero and increase to the initial value near the end of the mission. For example, when the initial inclination is selected at 3.5 deg, the inclination will decrease to zero and then increase to 3.5 deg in approximately 7 yr, depending on the epoch. The corresponding initial node is around 270 deg, which must be determined by iteration. The long-term inclination variations will be discussed later in Sec. 10.10.

For missions like DSCS III, north-south stationkeeping is required to keep inclination variation small, less than 0.1 deg. The inclination is controlled by applying ΔV normal to the orbit plane at desirable locations. The nominal locations are at the ascending or descending node. However, for eccentric orbits, the location may be at other points on the orbit to minimize ΔV. The amount of ΔV is determined from the desired inclination correction with the following equation:

$$\Delta V = 2V \sin \frac{\Delta \Theta}{2} \qquad (10.9)$$

where V is the circumferential velocity at the point of correction, and $\Delta \Theta$ is the total plane change given by

$$\Delta \Theta = \cos^{-1}[\sin i_1 \sin i_2 \cos(\Omega_2 - \Omega_1) + \cos i_1 \cos i_2] \qquad (10.10)$$

In Eq. (10.10), i and Ω are the inclination and right ascension of the ascending node of the orbit plane, respectively. The subscripts 1 and 2 denote conditions before and after the inclination correction.

For a 0.1-deg north-south stationkeeping tolerance, the ΔV required is of the order of 50 m/s, with four to five inclination control maneuvers per year.

10.7   Solar-Radiation-Pressure Effects

The solar-radiation pressure is induced by the light energy (photons) radiated from the sun. At one A.U. (astronomical unit), the solar-radiation-pressure constant P_0 is 4.7×10^{-5} dyne/cm^2. The perturbing acceleration of an Earth satellite due to solar-radiation-pressure effects can be computed by means of the following equation:

$$\frac{a_p}{g_0} = 0.97 \times 10^{-7}(1 + \beta) \left(\frac{A}{W} \right) \left(\frac{a_\odot}{r_\odot} \right)^2 \qquad (10.11)$$

where a_p is the magnitude of solar-radiation-pressure acceleration, g_0 is the Earth gravitational acceleration at sea level (32.2 ft/s^2 or 9.8 m/s^2), and $a_\odot / r_\odot$ is approximately 1 for near-Earth orbits. Also,

$$\beta = \text{optical reflection constant}$$

where

$\begin{cases} \beta = 1 \text{ total reflection (mirror)} \\ \beta = 0 \text{ total reception (blackbody)} \\ \beta = -1 \text{ total transmission (transparent)} \end{cases}$

A = effective satellite projected area, ft^2

W = total satellite weight, lb

$r_\odot, a_\odot$ = semimajor axis and radius of the sun's orbit around Earth

The direction of the acceleration a_p is perpendicular to the effective area A, which may or may not be perpendicular to the sun's ray. For a sphere or a flat plate perpendicular to the sun's ray, the normalized components of solar-radiation acceleration can be expressed by the following equations:

$$
\begin{Bmatrix} F_r \\ F_s \end{Bmatrix} = \cos^2 \frac{i}{2} \cos^2 \frac{\varepsilon}{2} \begin{Bmatrix} \cos \\ \sin \end{Bmatrix} (\lambda_\odot - u - \Omega)
$$

$$
- \sin^2 \frac{i}{2} \sin^2 \frac{\varepsilon}{2} \begin{Bmatrix} \cos \\ \sin \end{Bmatrix} (\lambda_\odot - u + \Omega)
$$

$$
- \frac{1}{2} \sin i \sin \varepsilon \left[\begin{Bmatrix} \cos \\ \sin \end{Bmatrix} (\lambda_\odot - u) - \begin{Bmatrix} \cos \\ \sin \end{Bmatrix} (-\lambda_\odot - u) \right] \quad (10.12)
$$

$$
- \sin^2 \frac{i}{2} \cos^2 \frac{\varepsilon}{2} \begin{Bmatrix} \cos \\ \sin \end{Bmatrix} (-\lambda_\odot - u + \Omega)
$$

$$
- \cos^2 \frac{i}{2} \sin^2 \frac{\varepsilon}{2} \begin{Bmatrix} \cos \\ \sin \end{Bmatrix} (-\lambda_\odot - u - \Omega)
$$

$$
F_w = \sin i \cos^2 \frac{\varepsilon}{2} \sin(\lambda_\odot - \Omega)
$$

$$
- \sin i \sin^2 \frac{\varepsilon}{2} \sin(\lambda_\odot + \Omega) - \cos i \sin \varepsilon \sin \lambda_\odot
$$

where F_r, F_s, and F_w are components of acceleration along the satellite orbit radius vector, perpendicular to F_r in the orbital plane, and along the orbit normal, respectively. The parameters, i, u, and Ω are orbit parameters defined in Chapters 3 and 9.

Here ε denotes the obliquity of the ecliptic, and $\lambda_\odot$, the ecliptic longitude of the sun. The quantities ε, $\lambda_\odot$, and $a_\odot/r_\odot$ and can be computed with sufficient accuracy from the expressions (see *Explanatory Supplement to the Astronomical Ephemeris*, 1961, p. 98).

$$
d = \text{MJD} - 15019.5
$$
$$
\varepsilon = 23°.44
$$
$$
M_\odot = 358°.48 + 0°.98560027d
$$
$$
\lambda_\odot = 279°.70 + 0°.9856473d + 1°.92 \sin M_\odot
$$
$$
a_\odot/r_\odot = [1 + 0.01672 \cos(M_\odot + 1°92 \sin M_\odot)]/0.99972
$$

where MJD is the modified Julian day. The modified Julian day = Julian day − 2400000.5.

By substituting the components F_r, F_s, and F_w into the equation of (9.21) of Chapter 9, one would obtain the variations of orbit elements due to radiation-pressure effects. After examining the dominant terms in the equations of variation, the most significant effect is on orbit eccentricity. For geosynchronous orbits, the dominant term in the equation of variation for eccentricity is $\sin(\lambda_\odot - \omega - \Omega)$. This term suggests that the eccentricity has a long-period (of the year) varia-tion, with the magnitude depending on β, A/W and initial conditions. The same variation appears in the argument of perigee. By properly choosing a unique set of initial conditions, the long-period eccentricity variation will disappear, and a constant eccentricity, sometimes called the "forced eccentricity" or "resonance eccentricity," will exist as a result of the resonance property discussed in Ref. 11. Therefore, for missions requiring very small eccentricity (or circular orbit), the design of the spacecraft should take into account reducing A/W. Regular eccentricity control maneuvers may be necessary to keep the eccentricity value small in the presence of significant solar-radiation-pressure effects.

An Algorithm to Compute Eccentricity Variations of Geosynchronous Orbits

For a geosynchronous orbit, the eccentricity variations due to solar-radiation pressure can be expressed by a simple relation in closed form[11] as follows:

$$e = \left[\left(\frac{g}{z} \right)^2 - 2\rho \left(\frac{g}{z} \right) \cos\theta + \rho^2 \right]^{\frac{1}{2}} \tag{10.13}$$

where

$$\rho = \left[e_0^2 - \left(\frac{2g}{z} \right) e_0 \cos\phi_0 + \left(\frac{g}{z} \right)^2 \right]^{\frac{1}{2}} \tag{10.14}$$

$$\theta_0 = \sin^{-1} \left(\frac{e_0}{\rho} \sin\phi_0 \right) \tag{10.15}$$

$$\phi_0 = \lambda_\theta - \omega_0 - \Omega_0 \tag{10.16}$$

$$g = \frac{3}{2} \frac{(1+\beta)(A/M)P}{V} \left(1 - e_0^2 \right)^{\frac{1}{2}} \cos^2 \left(\frac{\varepsilon}{2} \right) \tag{10.17}$$

and

e_0 = initial orbit eccentricity
A/M = cross-sectional area/mass ratio of the satellite
P = 4.65×10^{-6} N/m^2
$\lambda_\odot$ = ecliptic longitude of the sun at epoch
ω_0 = argument of perigee at epoch
Ω_0 = right ascension of ascending node at epoch
z = $\dot{\lambda}_\odot - \dot{\Omega}$
Θ = $\Theta_0 + z(t - t_0)$
V = velocity of satellite
ε = obliquity of the ecliptic ≈ 23.5 deg
$\dot{\lambda}_\odot$ = rate of the ecliptic longitude of the sun (≈ 1 deg/day)
$\dot{\Omega}$ = nodal regression rate

From the point of view of physical phenomena, the long-term eccentricity variation may be explained by the following analogy with energy addition and subtraction. Consider a circular orbit with counterclockwise motion, as shown in

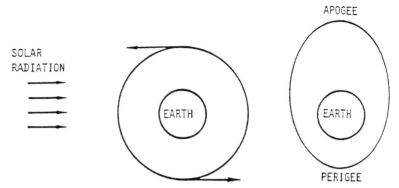

Fig. 10.9   Solar-radiation-pressure effect on an Earth satellite.

the center of Fig. 10.9. The solar radiation coming from the left is equivalent to adding a ΔV at the lower end of the orbit and subtracting the same ΔV at the upper end of the orbit. As a result, the orbit gradually evolves into an elliptical orbit like the one at the right of Fig. 10.9. Six months later, the sun is at the opposite side of the orbit, and the ΔV reverses sign at both ends of the elliptical orbit. Hence, the orbit will be circularized gradually because of the radiation-pressure effect.

For interplanetary missions, the solar-radiation-pressure effect was once considered by engineers at JPL as a means to navigate spacecraft for inner planet encounters. This concept is called "solar sailing." By properly orienting the "sails" of the spacecraft, a desired component along the velocity vector may be achieved. Thus, solar energy can be added or subtracted to the orbit, as shown by Fig. 10.10.

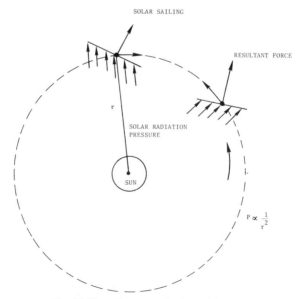

Fig. 10.10   Concept of solar sailing.

Study results have demonstrated that the specially guided energy transfer can, in fact, be used to navigate the spacecraft to its target planet. However, two technical problems hinder the realization of this concept. First, the solar-radiation pressure is inversely proportional to the square of sun-spacecraft distance. Thus, this technique is effective when the spacecraft is closer to the sun. The heat of the radiation poses great difficulty in manufacturing a spacecraft that can resist the severe heat. Second, the large area and light weight make it another very difficult task to design and manufacture such a spacecraft.

10.8 Atmospheric Drag Effects

The Earth's atmosphere produces drag forces that retard a satellite's motion and alter the orbit shape. A low-altitude satellite will eventually be slowed until it spirals in. The rate of orbital decay depends on atmospheric density; this varies with time and geographic position and is not precisely known. Orbital parameters such as height, and the ballistic coefficient, which takes a spacecraft shape and weight into account, are also important factors that must be considered. The effect of atmospheric drag on a satellite orbit is illustrated in Fig. 10.11.

The ballistic drag coefficient $W/C_D A$ is the quotient of a mass quantity divided by a drag quantity. The more "massive" the object and/or the smaller its drag, the greater will be the value of the ballistic coefficient and, at the same time the less the object will be slowed as a result of its passage through the atmosphere.

The value of the drag coefficient C_D depends on the shape of the vehicle, its attitude with respect to the velocity vector, and whether it is spinning, tumbling, or is stabilized. Above 200-km altitude, the drag coefficient varies from about 2.2 for a sphere to about 3.0 for a cylinder, with other shapes being somewhere in between. Exact values of C_D are best determined by actual flight test. A value of 2.2 will yield a conservative result.

When the orbit altitude is less than 1000 km, atmospheric drag effect should be considered in long-term predictions. The equation for computing drag acceleration is

$$a_D = -\frac{1}{2}\rho g_0 V^2 \frac{C_D A}{W} i_v \qquad (10.18)$$

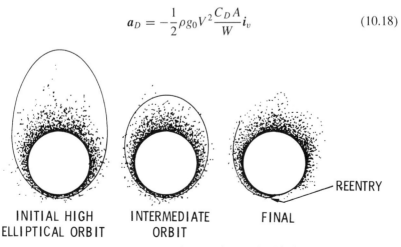

INITIAL HIGH   INTERMEDIATE   FINAL
ELLIPTICAL ORBIT   ORBIT

Fig. 10.11   Aerodynamic drag–induced orbit decay.

where

a_D = atmosphere drag acceleration vector

ρ = atmosphere density

g_0 = Earth gravitation acceleration at sea level

V = velocity of the satellite

A = satellite effective (projected) area

W = satellite weight at sea level

i_v = unit vector of the satellite velocity

and C_D is the dimensionless drag coefficient of the spacecraft. The difficult part of the drag prediction is the modeling of the density of the atmosphere, which is a function of both altitude and time. The atmosphere density at a given altitude increases with solar flux, which has an 11-yr cycle. The air density also exhibits a day-to-night rhythm, reaching a maximum about 2 h after midday and a minimum between midnight and dawn. According to King-Hele,[13] at a height of 600 km, the maximum daytime density may be as large as eight times greater than the nighttime minimum. Another variation in the air density is the 27-day cycle caused by the extreme ultraviolet (UV) radiation.

Numerous mathematical models have been developed by many researchers since the late 1950s. The most commonly used models are the following:

1) The ARDC 1959 model, which is a static model and is based in part on density data inferred from early satellite observation.

2) The U.S. Standard Atmosphere of 1962, which was designed to represent an idealized, middle-latitude, year-round mean over the range of solar activity between sunspot minima and sunspot maxima. This model is also a static model, and the density values are consistently less than the ARDC 1959 model at various altitudes.

3) The dynamic Jacchia 1964 model, which accounts for the diurnal, 27-day, and 11-yr cycles. The Jacchia 1964 model has been widely used for accurate drag effect predictions. The values of density according to the Jacchia 1964 model for three representative periods of solar activity are shown in Fig. 10.12.

4) The Jacchia 1971 model, widely used for orbit decay predictions, is a dynamic model with improved accuracy.

5) The MSIS90E density model, which is considered the most accurate density model, contains a density profile extending all the way to the Earth's surface.

These three conditions correspond approximately to $F_{10.7}$ (an index used in specifying solar activity) values of 220, 150, and 70 for active, average, and quiet solar periods, respectively. In Fig. 10.12, the density values of the ARDC 1959 and U.S. Standard 1962 models are also shown for comparison.

The drag effect on satellite orbits takes place through energy dissipation. When an orbit loses energy, its semimajor axis decreases. For an orbit with large eccentricity, the drag effect would first circularize the orbit by gradually lowering its apogee and, then, the radius of a circular orbit would continue to decrease until the satellite crashes on the Earth's surface. For circular orbits, the orbit decay rate can be computed by the following equation:

$$\frac{da}{dt} = -\sqrt{\mu a}\left(\frac{\rho g_0}{B}\right) \tag{10.19}$$

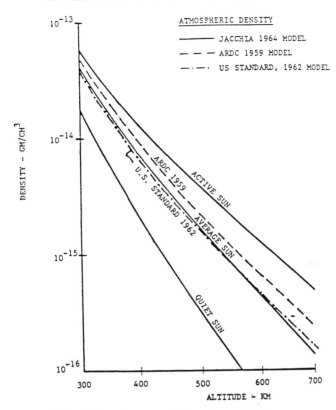

Fig. 10.12   Atmospheric density vs altitudes.

where μ is the Earth gravitational constant, a the semimajor axis, and B the ballistic coefficient $(W/C_D A)$.

Recently, the concept of using drag as a means to save propellant in orbit-transfer maneuvers has received some attention. The transfer from a high-altitude to a low-altitude orbit may be achieved by first lowering the perigee into the atmosphere and then letting the drag gradually reduce the apogee to the desired value. Finally, the perigee may be raised to a higher value if needed. If the spacecraft may be so oriented as to have a lifting surface, the atmospheric pressure can be utilized to make plane changes for highly eccentric orbits. The rotation of the atmosphere would also yield a small component normal to the orbits with high inclination. This component may change the inclination to 0.1 deg in 100 days. This concept is sometimes called the "aeroassisted orbit transfer."

In Fig. 10.13, the approximate lifetime of a circular orbiting satellite is shown in days as a function of its altitude. Several representative values of the ballistic coefficient are given. It should be noted that, for a first approximation, the lifetime is a linear function of the ballistic coefficient $W/C_D A$.

The curves in Fig. 10.14 represent orbits of differing eccentricities. The numbers alongside the curves give approximate period times, in minutes, at that point on the curve.

Perigee altitude is shown on the abscissa, and the ordinate is an odd quantity, the number of revolutions divided by the ballistic coefficient. To find the lifetime

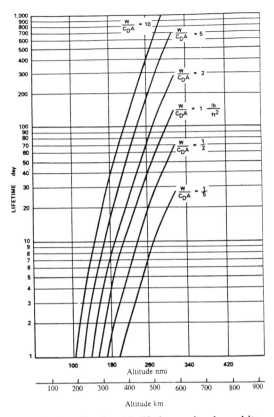

Fig. 10.13   Earth orbit lifetimes: circular orbits.

from this ordinate, the calculated value of the ballistic coefficient is multiplied by the orbital period in minutes. The results in Figs. 10.13 and 10.14 are very approximate as they are based on a static atmosphere.

It should be noted that, because of the atmospheric density changes that result from the 11-yr solar cycle, large variations in orbital lifetime can occur during the course of several years. As solar activity (solar flares and sunspots) increases the effective density, the height of the Earth's upper atmosphere is increased: as a result, lifetime may be less than that shown in Figs. 10.13 and 10.14. The lifetimes determined from the figures should be taken as mean values. Of course, if solar activity decreases, upper-atmosphere densities will be less, and lifetime longer, than shown.

10.9   Tidal Friction Effects

The tidal friction effects can be explained by the Earth-moon system, as illustrated in Fig. 10.15. The bulge is caused by the gravity pull of the moon, and the tidal friction exerts a net torque on Earth. The phase delay ψ occurs because Earth is spinning at a faster rate than the rotation rate of the moon around Earth. As a result of energy dissipation through friction and the conservation of angular

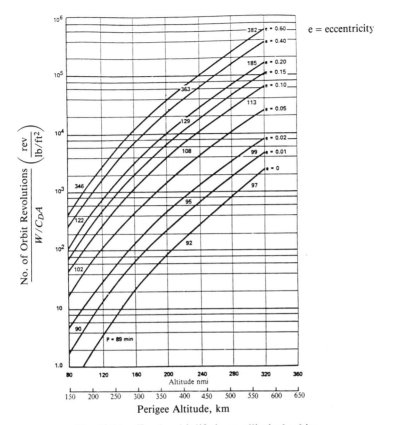

Fig. 10.14   Earth orbit lifetimes: elliptical orbits.

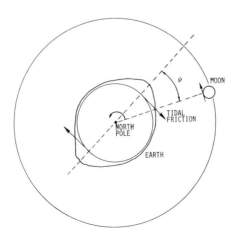

Fig. 10.15   The Earth-moon tidal friction mechanism.

momentum of the Earth-moon system, the Earth spinning rate is slowly deceler-
ating, and the moon is slowly moving away from Earth. This mechanism is true
for other outer planets with massive satellites.

The four massive satellites of Jupiter have the common features of most inner
satellites of the solar system. These features can be summarized as follows:

1) Small inclination to planet's equatorial plane.
2) Small eccentricity.
3) Orbit-orbit resonance.

Extensive studies have been made in recent years to explain the earlier-listed
phenomena. According to Goldreich,[14] dynamical considerations strongly favor
low inclinations for close satellites of an oblate precessing planet. The small
eccentricity and orbit-orbit resonance are believed to be consequences of tidal
dissipation. For the evolution of small eccentricities, a mechanism was proposed
by Urey et al.[15] in the form of tidal working in the satellite due to tides raised
by the planet. This mechanism can be explained by a simple relation between
eccentricity and energy of the satellite orbit.

$$e = \left(1 + \frac{2EL^2}{M_s^3 M_p^2} G \right)^{\frac{1}{2}} \tag{10.20}$$

where E is the energy of the orbit, L is the angular momentum, and M_p and M_s
are the planet and satellite masses, respectively. G is the universal gravitational
constant. If the satellite is not spinning, the tide raised on it can produce only a
radial perturbation force. This means that L is not changed by the tide. Since any
energy dissipation in the satellite decreases E and since we have $E < 0, 0 < e < 1$,
and L constant, we find that e is decreasing also. This process will not stop until
$e = 0$ because the height of the tide will vary with the oscillation in distance
between the satellite and planet. Later, Goldreich and Soter[16] examined such a
mechanism, with reasonable values of the tidal energy dissipation function Q, and
suggested that tides raised on satellites are of great significance in the evaluation
of the eccentricities of these satellites.

A tidal origin of commensurable satellite mean motions was first proposed by
Goldreich in 1964,[17] with the following evidence:

1) There are more examples of resonances than can be explained by random
distribution of satellite orbital elements.
2) Tides raised on a planet by a satellite tend to vary according to the satellite's
orbital period at a rate dependent on the satellite's mass and distance.
3) Mutual gravitational interaction is strong enough to maintain the commensu-
rability of resonant satellites' periods, even against the upsetting influence of the
tides.

Later, in 1972, Greenberg[18] gave a realistic model (Titan-Hyperion) involving
mutual gravitation and tidal dissipation that provides a detailed explanation for
satellite orbit-orbit resonance capture. The conclusion of his study is that once
resonant commensurabilities are reached, the angular momentum transferred as
a result of tidal friction on the planet should be distributed in such a way that
commensurabilities among the four inner satellites are maintained.

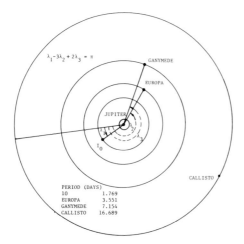

Fig. 10.16   The Galilean satellites of Jupiter.

Although tidal dissipation has been shown to be responsible for small eccentricities and stable commensurabilities, the magnitude of the tidal force exerted on the Galilean satellites is several orders of magnitude smaller[17] than other perturbing forces such as oblateness and mutual and solar attractions. This is true for all the natural and artificial satellites of the solar system.

It is interesting to point out that, within observational accuracy, the mean motions of the inner three Galilean satellites obey the relation $n_1 - 3n_2 + 2n_3 = 0$. The mean longitudes of the satellite satisfy the liberation equation $\lambda_1 - 3\lambda_2 + 2\lambda_3 = 180$, as illustrated in Fig. 10.16. The four massive satellites were discovered by Galileo in 1610 after he invented the telescope.

10.10   Long-Term Inclination Variations

The long-term orbit inclination variation has been one of the important elements in orbit selection and mission design. Inclination control maneuvers require significantly more propellant than the longitude or in-track orbit maintenance maneuvers. Therefore, a good understanding of the inclination perturbations and a proper selection of an initial orbit can often avoid costly ΔV expenditure and reduce the payload.

The long-term inclination perturbations are caused primarily by the sun/moon attractions. In Ref. 7, the averaged equations of variation for third-body perturbations are given in computer-assisted series expansion form, with orbit eccentricity in closed form. The equations for inclination and right ascension of ascending node variations can be reduced for near-circular orbits as

$$\frac{di}{dt} = -\frac{3}{4}\frac{y}{\sin i}\text{IPO}$$

$$\frac{d\Omega}{dt} = \frac{3}{4}y\left(\frac{\cos i}{\sin i}\text{IPS} - \text{IPC}\right)$$

(10.21)

where

$$y = n'^2/n(a'/r')^3 R_m$$

n = mean motion of the satellite orbit

n' = mean motion of the perturbing body (sun or moon)

R_m = mass ratio, $R_m = 1$ for solar perturbation, $= 1/82.3$ for lunar perturbation

= mass of the perturbing body/(mass of the Earth + mass of the perturbing body)

and IPO, IPS, and IPC are computer-generated series listed here:

$$\text{IPO} = +(-1^*\text{SI}^*\text{CI}^*\text{SI3}^*\text{CI3})^*\text{SIN(D)}$$

$$+(-1/2^*\text{SI}^{**}2^*\text{SI3}^{**}2)^*\text{SIN}(2^*\text{D})$$

$$+(+1/2^*\text{SI}^{**}2 + 1/2^*\text{SI}^{**}2^*\text{CI3} - 1/4^*\text{SI}^{**}2^*\text{SI3}^{**}2)^*\text{SIN}(2^*\text{L} - 2^*\text{D})$$

$$+(-1/2^*\text{SI}^*\text{CI}^*\text{SI3} - 1/2^*\text{SI}^*\text{CI}^*\text{SI3}^*\text{CI3})^*\text{SIN}(2^*\text{L} - \text{D})$$

$$+(-1/2^*\text{SI}^*\text{CI}^*\text{SI3} + 1/2^*\text{SI}^*\text{CI}^*\text{SI3}^*\text{CI3})^*\text{SIN}(2^*\text{L} + \text{D})$$

$$+(-1/2^*\text{SI}^{**}2 + 1/2^*\text{SI}^{**}2^*\text{CI3} + 1/4^*\text{SI}^{**}2^*\text{SI3}^{**}2)^*\text{SIN}(2^*\text{L} + 2^*\text{D})$$

$$\text{IPS} = +(+1^*\text{SI}^*\text{SI3}^{**}2)$$

$$+(+1^*\text{CI}^*\text{SI3}^*\text{CI3})^*\text{COS(D)}$$

$$+(-1/2^*\text{CI}^*\text{SI3} - 1/2^*\text{CI}^*\text{SI3}^*\text{CI3})^*\text{COS}(2^*\text{L} - \text{D})$$

$$+(-1^*\text{SI}^*\text{SI3}^{**}2)^*\text{COS}(2^*\text{L})$$

$$+(+1/2^*\text{CI}^*\text{SI3} - 1/2^*\text{CI}^*\text{SI3}^*\text{CI3})^*\text{COS}(2^*\text{L} + \text{D})$$

$$\text{IPC} = +(+1^*\text{CI} - 1/2^*\text{CI}^*\text{SI3}^{**}2)$$

$$+(+1^*\text{SI}^*\text{SI3}^*\text{CI3})^*\text{COS(D)}$$

$$+(-1/2^*\text{CI}^*\text{SI3}^{**}2)^*\text{COS}(2^*\text{D})$$

$$+(-1/2^*\text{CI} - 1/2^*\text{CI}^*\text{CI3} + 1/4^*\text{CI}^*\text{SI3}^{**}2)^*\text{COS}(2^*\text{L} - 2^*\text{D})$$

$$+(-1/2^*\text{SI}^*\text{SI3} - 1/2^*\text{SI}^*\text{SI3}^*\text{C13})^*\text{COS}(2^*\text{L} - \text{D})$$

$$+(+1/2^*\text{CI}^*\text{SI3}^{**}2)^*\text{COS}(2^*\text{L})$$

$$+(+1/2^*\text{SI}^*\text{SI3} - 1/2^*\text{SI}^*\text{SI3}^*\text{CI3})^*\text{COS}(2^*\text{L} + \text{D})$$

$$+(-1/2^*\text{CI} + 1/2^*\text{CI}^*\text{CI3} + 1/4^*\text{CI}^*\text{SI3}^{**}2)^*\text{COS}(2^*\text{L} + 2^*\text{D})$$

where $\text{SI} = \sin i$, $\text{CI} = \cos i$, $\text{SI3} = \sin i'$, $\text{CI3} = \cos i'$, $L = U'$, $D = \Omega - \Omega'$ ( )$'$ = elements of the third body (sun or moon).

The equation set (10.21) can be further reduced by eliminating those intermediate-period terms (terms with U' as argument) and combining the effect due to J_2 to

one (or more than one) of the three constraints just listed will be violated? What will be the new rates of node and argument of perigee?

Note: Consider only the perturbation due to J_2, $J_2 = 0.0010826$, $\mu = 0.39860047 \times 10^{15}$ m^3/s^2, R_e (Earth equatorial radius) = 6378140 m.

10.3. Determine the semimajor axis of an Earth satellite orbit with eccentricity equal to 0.17 and $d\omega/dt = 0$ (critical inclination, $i = 63.4$ deg or 180 deg $- 63.4$ deg), and the orbit is sun-synchronous.

10.4. a) Write down the equations of motion in relative form with perturbations a_p. If the moon is the perturbing body, write down $a_p =?$

b) Briefly explain the methods of solution to these equations of motion. If high-accuracy orbit prediction or orbit determination is required, what method should be used and why? If long-term orbit integration is required and the emphasis is on computation speed rather than accuracy, what method should be used?

c) For low-altitude orbits (185–740 km), what perturbations are important? For high-altitude orbits (GPS and geosynchronous altitudes), what perturbations are important and why?

10.5. The potential function for the primary body with J_2 effect is

$$\phi = \frac{u}{r}\left[1 + \frac{1}{2}J_2\left(\frac{a_e}{r}\right)^2(1 - 3w^2)\right]$$

where

$\mu = k^2m$ gravitational constant

$a_e =$ Earth equatorial radius

$r =$ geocentric distance to satellite $r = \sqrt{x^2 + y^2 + z^2}$

$w = z/r$

Derive the equations of motion from the preceding potential in the Cartesian coordinates. (Hint: $d^2x/dt^2 = \partial\phi/\partial x, x \rightarrow y, z$.)

10.6. Compute and plot the time history of the right ascension of ascending node Ω of an Earth orbit with the following initial conditions:

$a_0 = 3700$ n.mi.

$e_0 = 0$

$i_0 = 30$ deg

$\Omega_0 = 45$ deg

$\omega_0 = 50$ deg

$M_0 = 100$ deg

($\mu = 62750.1633$ n.mi.^{3}/s^2, Earth equatorial radius = 3444 n.mi.)

Include only the J_2 ($J_2 = 0.0010826$) effect. Plot the Ω variations as a function of time (min) up to 300 min. After you have plotted them, you will notice that the

variations are a combination of secular and short-period variations. Which one of the two is more important?
(Hint: Find the equation $d\Omega/dt = -n(3/2)J_2(a_e/p)^2 \cos i\{1 + 3e\cos M + \cdots\}$ in the class notes, and analytically integrate the equation by ignoring all the terms having e as coefficient. Assume that $n, a, i,$ and ω are constant over the 300-min interval.)

10.7. Compute maximum orbital eccentricity for a 1000-kg synchronous equatorial satellite with a projected area $A = 10$ m^2 resulting from solar-radiation pressure. Assume total reflectivity for the surface $(\beta = 1)$. What if the area is 10^4 m^2 (balloon)?

10.8. What is the lifetime of a 1000-lb (454.5-kg) spherical satellite of 100-ft^2 area in a low-altitude 160×360 km (86.5×194.6 n.mi.) Earth orbit?

Selected Solutions

10.1. $\dot{\Omega} = -4.0, 0, -3.6$ deg/day
$\dot{\omega} = 1.0, -4.0, 0$ deg/day
(body), (90 deg), (63.4) deg

10.2. $i = 116.6$ deg
$\alpha = 10{,}187.6$ km

10.3. $a = 9981.25$ km

10.7. $e_{\max} = 0.00043, 0.43$

10.8. 1.9 days

11.1 Launch Window Considerations

Launch Azimuth

The launch of a satellite into an orbit with inclination i requires launching in an azimuth direction Az defined by the formula

$$\sin Az = \frac{\cos i}{\cos \phi_0} \qquad (11.1)$$

where ϕ_0 is the launch site latitude. For a due east launch ($Az = 90$ deg), $i = \phi_0$, or the orbital inclination is equal to the launch site latitude. Differentiating Eq. (11.1) and solving for di, one obtains

$$(di) = \frac{-\cos \phi_0 \cos Az \, dAz}{\sin i} \qquad (11.2)$$

which shows the sensitivity of inclination error di due to an error in azimuth dAz. For range safety reasons, the azimuth may be constrained between specified limits, as is shown in Fig. 11.1, for example.

In-Plane or Out-of-Plane Ascent Using a Launch Azimuth Constrained to the Nominal Value

The most efficient launch is an in-plane launch ascent characterized by a wait on the ground until the launch site lies in the mission orbit plane, which is fixed in inertial space. No plane change maneuver is then required unless the orbit inclination is less than the launch site latitude or the launch azimuth constraint is such that a "dog-leg" maneuver is needed. A direct ascent to a parking or mission orbit may be made with or without a *phasing orbit* to place the satellite in a specified geographic location. The phasing problem where the parking and mission orbits are not coplanar may involve excessive waiting times, which can be somewhat reduced by "lofted" or other trajectories.

The out-of-plane launch is a form of direct ascent when the phasing problem can be solved by a wait on the ground. The launch window geometry is shown in Fig. 11.2.

The launch window can be defined as the time,

$$\text{L.W.} = 2\Delta\Omega/\omega_e \qquad (11.3)$$

where $\Delta\Omega = \omega_e t$ is the maximum permissible nodal increment consistent with the plane change angle ε required to transfer from the launch orbit plane to the mission orbit plane. The latter can be determined from the equation

$$\varepsilon = \cos^{-1}(\cos^2 i + \sin^2 i \cos \Delta\Omega) \qquad (11.4)$$

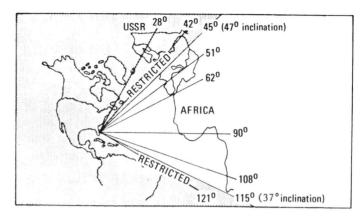

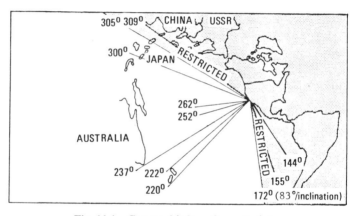

Fig. 11.1   Geographic launch constraints.

where i is the inclination of the orbit and the corresponding launch azimuth is constrained to the nominal value provided by Eq. (11.1). The longitudinal separation $\Delta\Omega$ is the same at all latitudes.

In general, a graph of ε vs time is as shown in Fig. 11.3, where the plane change capability ε of the launch vehicle defines the launch window.

Launch Azimuth Constrained to Off-Nominal Value

If the launch azimuth is constrained to an off-nominal value, then Fig. 11.4 depicts the launch geometry at the time $t = 0$ the launch site lies in the mission orbit plane. The launch geometry at some later time, $t > 0$, is shown in Fig. 11.5. Applying spherical trigonometry to the triangle bounded by the launch orbit, mission orbit, and equator results in the equation

$$\cos \varepsilon = \cos i \cos i' + \sin i \sin i' \cos(\omega_e t + \Delta\Omega_0) \qquad (11.5)$$

where $\Delta\Omega_0$ is the difference in right ascension of the ascending node of the launch and mission orbits at $t = 0$.

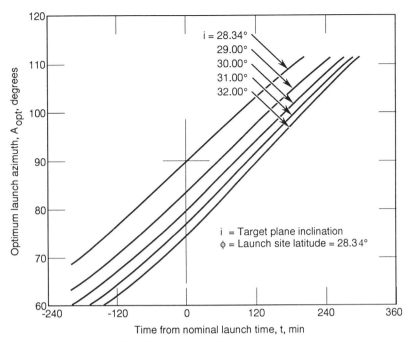

Fig. 11.9   Optimum launch azimuth.

duration of the mission. This angle, conventionally known as the beta angle β, is illustrated in Fig. 11.10. Since β is the complement of the angle between the sun vector $\hat{s}$ and the positive normal to the orbit $\hat{n}$, it follows from their scalar product that

$$\beta = \sin^{-1}(\hat{s} \cdot \hat{n}) \qquad (11.9)$$

and ultimately that[1]

$$\beta = \sin^{-1}[\cos \delta_S \sin i \sin(\Omega - \text{RATS}) + \sin \delta_S \cos i] \qquad (11.10)$$

where β is defined to lie in the range from -90 to $+90$ deg.

Equation (11.10) reveals that beta angle depends on solar declination δ_S, orbit inclination i, and the difference in right ascensions of the true sun and the ascending node $(\Omega - \text{RATS})$. The first of these quantities, δ_S, depends on the date during the mission. The second quantity, i, is essentially constant during the mission. The last quantity $(\Omega - \text{RATS})$ changes because of nodal regression (induced by Earth's oblateness perturbations, as described in Chapter 8) and seasonal variation in the right ascension of the true sun.

In light of the variability of the terms on the right side of Eq. (11.10), it is clear that the beta angle cannot be held constant throughout a mission. However, it is generally possible to select conditions at the start of the mission so that the beta angle will stay within some prescribed tolerable range of values for that portion of the mission during which β is essential to performance.

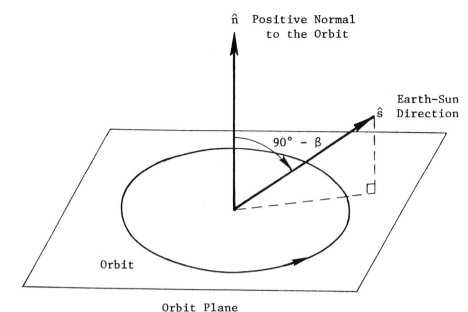

Fig. 11.10   Definition of sun-orbit angle, beta.

Earth Eclipsing of a Circular Orbit

It may be important to determine those occasions during its mission when a satellite is eclipsed by the Earth. Such eclipsing occurs whenever the satellite passes through the Earth's shadow, which is assumed cylindrical in this discussion.* Figure 11.11 shows that the Earth's shadow intersects the orbital sphere of a satellite at altitude h in a minor circle whose Earth-central-angular radius is β^*, where[1]

$$\beta^* = \sin^{-1}[R/(R + h)], \qquad 0 \deg \le \beta^* \le 90 \deg \qquad (11.11)$$

View A–A in Fig. 11.11 reveals that the orbit intersects the perimeter of the shadow circle at points E_1 and E_2. Note that the length of the eclipsed orbital arc E_1E_2 is just twice arc CE_1, where C is the point on the orbit of closest approach to the shadow axis A. That is, the length of arc AC is just the magnitude of β. Hence, it follows from the right spherical triangle ACE_1 that

$$\Delta u = \cos^{-1}(\cos \beta^* / \cos \beta) \qquad (11.12)$$

When Eqs. (11.11) and (11.12) are combined, the eclipsed fraction of the circular orbit is found to be[1]

$$f_E = \frac{2\Delta u}{2\pi} = \frac{1}{\pi} \cos^{-1}\left[\frac{\sqrt{h^2 + 2Rh}}{(R + h)\cos \beta} \right] \qquad (11.13)$$

---

*Such an assumption is valid at low satellite altitudes, where there is no appreciable difference between the umbral and penumbral regions of total and partial eclipsing, respectively.

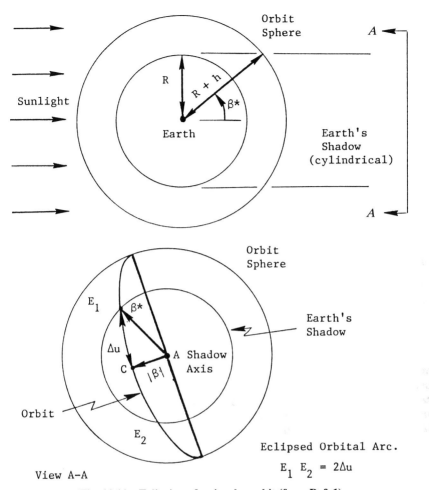

Fig. 11.11   Eclipsing of a circular orbit (from Ref. 1).

Figure 11.12 shows the variation of eclipse fraction with β-angle magnitude for several low-altitude circular orbits. Note that the eclipse fraction diminishes with an increase in either altitude or magnitude of β. One should also observe that eclipsing can occur only if $|\beta|$ is less than the critical angle β^* for the given orbit altitude. For example, a satellite in a 200-n.mi.- altitude circular orbit will be eclipsed through some portion of its orbit if and only if $|\beta| < 70.93$ deg.

For a circular orbit, the time duration of the eclipse interval is directly proportional to its angular extent; in other words, the duration of the eclipse on a given satellite revolution is just the product of the eclipse fraction f_E and the orbital period. However, this is not true for eccentric orbits. The duration and angular extent of eclipsing of a satellite in an eccentric orbit depends on the eccentricity and the relative orientation of sun and perigee, as well as the sun-orbit angle β.

Figure 11.13 presents the Earth, its shadow, and the circular synchronous equatorial orbit at the time of the vernal equinox. The circular synchronous orbit radius

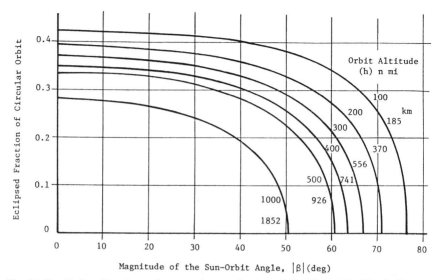

Fig. 11.12   Eclipse fraction vs beta-angle magnitude and circular orbit altitude (from Ref. 1).

of 6.61 Earth radii produces an orbit period of 24 sidereal hours. A satellite in this orbit revolves with an angular rate that exactly equals the angular rate of rotation of the Earth on its axis.

An Earth-centered inertial (ECI) coordinate system whose origin is at the center of the Earth is shown in this figure. The X axis points to the vernal equinox, the Z axis points to the celestial North Pole, and the Y axis forms a right-hand triad with X and Z.

In Fig. 11.13a, the sun is along the X axis, and the shadow axis intersects the synchronous equatorial orbit. The Y axis points 23.45 deg below the ecliptic plane.

This angular difference of 23.45 deg is important at the summer solstice (Fig. 11.13b) because the Earth casts a shadow that passes beneath the circular, synchronous equatorial orbit. And so no eclipsing occurs at this time.

Figure 11.14 shows the geosynchronous eclipse geometry. For geometrical purposes, consider the relative motion of the Earth's shadow as it moves along the ecliptic plane. At the distance of the synchronous orbit, the radius of the disk of the umbra is 8.44 deg. Therefore, 21.6 deg or 22 days before the autumnal equinox, the umbral disk becomes tangent to the equatorial plane. This is the beginning of the eclipse season because this is the first time a satellite in the synchronous equatorial orbit could experience an umbral eclipse, although it would be vanishingly short. At the time of the autumnal equinox, the center of the umbral disk would lie in the equatorial plane. This geometry produces a maximum time of 67.3 min.

Twenty-two days later, the eclipse season ends when the umbral disk is again tangent to the equatorial plane. And so, the eclipse season is 44 days, centered on the autumnal equinox. A similar eclipse season is centered on the vernal equinox.

Sun Synchronism of an Orbit

In Chapter 8, one of the principal perturbations of an orbit caused by Earth's oblateness was identified as nodal regression $\dot{\Omega}$. Recall that Eq. (8.3) showed that

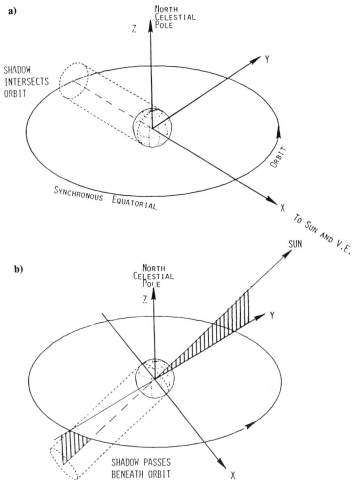

Fig. 11.13 Geosynchronous orbit geometry: a) vernal equinox shadow; b) summer solstice shadow.

$\dot{\Omega}$ depends as follows on orbit inclination i, average altitude h, and eccentricity e:

$$\dot{\Omega} = \frac{-9.9639}{(1 - e^2)^2} \left(\frac{R}{R + h} \right)^{3.5} \cos i \frac{\text{deg}}{\text{mean solar day}} \tag{11.14}$$

where

$h = (h_a + h_p)/2$

$e = (h_a - h_p)/(2R + h_a + h_p)$

and R = the equatorial radius of the Earth.

An orbit is said to be sun-synchronous if its line of nodes rotates eastward at exactly the orbital angular velocity of the mean sun. Since the mean sun

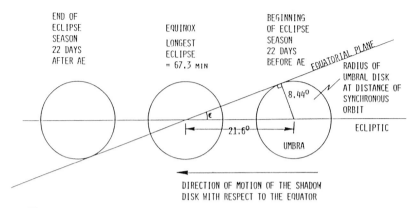

Fig. 11.14   Geosynchronous orbit eclipse at vernal or autumnal equinox.

moves uniformly eastward along the celestial equator through 360 deg in a tropical year (about 365.242 mean solar days), the required rate of nodal regression is 360/365.242, or 0.985647 deg/day. Substituting that numerical value in Eq. (11.14) and solving for the sun-synchronous inclination yields

$$i_{ss} = \cos^{-1}\left\{ -0.098922(1 - e^2)^2 \left(1 + \frac{h}{R}\right)^{3.5} \right\} \qquad (11.15)$$

Note that, since the cosine of i_{ss} is always negative, the inclination of a sun-synchronous orbit must be greater than 90 deg; that is, sun-synchronous orbits are necessarily retrograde.

For circular orbits, e is zero, and h is the constant altitude above a spherical Earth with radius R. In that case, the variation of sun-synchronous inclination with altitude is shown in Fig. 11.15. Note that sun synchronism is possible for retrograde circular orbits up to an altitude of about 3226 n.mi. or a radius of 6670 n.mi. (for $R = 3444$ n.mi.), at which point the orbit inclination reaches its greatest possible value, 180 deg. Figure 11.16 shows examples of noon-midnight and twilight orbits.

Launch Window to Satisfy Beta-Angle Constraints

The term *launch window*, as used here, will mean the time interval (or intervals) on a given date within which a satellite can be launched into a prescribed orbit with the subsequent satisfaction of sun-orbit angle constraints (upper and lower limits on β) for the specified mission duration.

If the mission duration were zero days, one would simply determine the time interval(s) on the given launch date within which the combination of solar declination δ_S, orbit inclination i, and sun-node orientation $(\alpha_S - \Omega)$ yields acceptable values of β, as shown in Eq. (11.10).

However, if the mission has any extent, the determination of launch window involves more effort. In such a case, one must account for the changes that occur in δ_S and in $(\alpha_S - \Omega)$ during the mission and select the reduced launch window accordingly. As was mentioned earlier, the seasonal change in δ_S is inevitable. However, the initial value of $(\alpha_S - \Omega)$ can easily be controlled by varying the

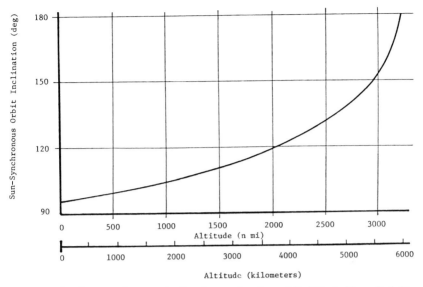

Fig. 11.15 Sun-synchronous inclination vs circular orbit altitude (from Ref. 1).

launch time; a 1-h delay in launch moves the line of nodes 15 deg eastward and thus reduces $(\alpha_S - \Omega)$ by 15 deg on any given launch date. Furthermore, the rate of change of $(\alpha_S - \Omega)$ can be adjusted by varying the orbit inclination or, to a lesser degree, the altitude. In particular, the average rate of change of $(\alpha_S - \Omega)$ can be made to vanish by employing a sun-synchronous combination of altitude and inclination. In this way, a preferred constant value of $(\alpha_S - \Omega)$ and an associated acceptable range of β values can often be maintained throughout the mission.

11.2 Time of Event Occurrence

If an event such as a perigee passage by a satellite occurs over a certain longitude λ, then the time t of the event can be specified unambiguously from the equation

$$\alpha_* = \text{GHA} + \lambda + \omega_e t \tag{11.16}$$
$$= \text{right ascension of the event}$$

where

GHA = Greenwich hour angle (at Greenwich midnight of date)
λ = longitude of the event
ω_e = 15.041067 deg/h = Earth rotation rate
t = time from Greenwich midnight (GMT)

The values of GHA can be found in standard reference works for each date. For example, on January 1, 2000, GHA = 99.97 deg. If the event is a perigee passage by a satellite, then,

$$\alpha_* = \Omega + \mu_{pp} \tag{11.17}$$

where

$$\mu_{pp} = \tan^{-1}(\tan \omega \cos i)$$

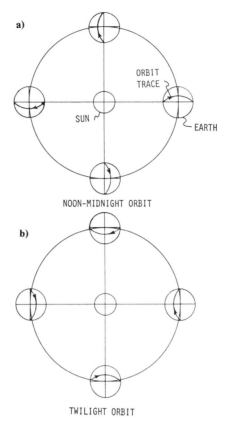

a)

ORBIT
TRACE

SUN

EARTH

NOON-MIDNIGHT ORBIT

b)

TWILIGHT ORBIT

Fig. 11.16   Sun-synchronous orbits: a) noon-midnight orbit ($\Omega - \lambda = 0$); b) twilight orbit ($\Omega - \lambda = 90$ deg).

and

$$\omega = \text{argument of perigee } (<180 \text{ deg})$$

The time of perigee passage t_{pp} is then given by

$$t_{pp} = (\Omega + \mu_{pp} - \text{GHA} - \lambda_{pp})/\omega_e \qquad (11.18)$$

where

$$\lambda_{pp} = \text{longitude of perigee passage}$$

11.3   Ground-Trace Considerations

General Characteristics

Circular figure eight, or "eggbeater," types of ground traces can be obtained by using satellites with 12- or 24-h periods at different inclinations to the equator. The use of several orbital planes equally spaced in node can result in several satellites moving in the same ground trace. Two examples of this are shown in

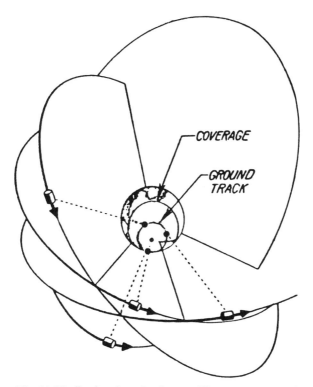

Fig. 11.17   Regional navigation satellite system concept.

Figs. 11.17 and 11.18, which show the circular and eggbeater types of ground traces with several satellites in each ground trace. Figure 11.17 is for a regional system with four satellites, whereas Fig. 11.18 is for a possible global navigation system employing twenty satellites.

Other orbital systems involving up to four 12-h satellites in each of six equally spaced orbit planes have been found useful for global navigation purposes, as in the global positioning system (GPS), for example.

One measure of performance of a navigation satellite system is the performance factor called *geometric dilution of precision* (GDOP), which is a measure of how satellite geometry degrades accuracy. The magnitude of the ranging errors to a minimum of four selected satellites, combined with the geometry of the satellites, determines the magnitude of the user position errors in the GPS navigation fix. The four "best" visible satellites are those with the lowest GDOP.[2] Thus,

$$GDOP = \sqrt{(PDOP)^2 + (TDOP)^2} \qquad (11.19)$$

where

  PDOP = ratio of radial error in user position 1 σ in three dimensions to range
       error 1 σ.
  TDOP = ratio of error 1 σ in the range equivalent of the user clock offset to
       range error 1 σ.

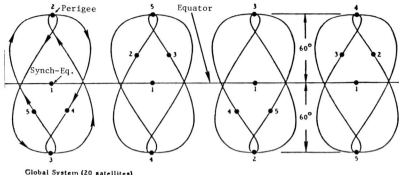

Global System (20 satellites)

4 satellites in each of 2 orbit planes 180° apart in longitude with perigee in southern hemisphere

4 satellites in each of 2 orbit planes 180° apart in longitude with perigee in northern hemisphere

4 satellites in synchronous equatorial orbit, 90° apart in longitude.

All inclined planes (i = 60°) with eccentricity e = .25 are separated 90° in longitude relative to each other.

Fig. 11.18   Global navigation satellite system concept.

Perturbation Effects

Earth's nonsphericity, solar-radiation pressure, and solar/lunar gravitational attraction tend to alter the satellites' orbital elements in time. An example of how the ground trace can change as a result of the Earth's gravitational harmonics is shown in Fig. 11.19, where the initial and a four-year ground trace are illustrated.

The results are for a $Q = 1$ eccentric ($e = 0.27$) orbit, whose period is $23^h55^m59.3^s$, with an initial inclination $i_0 = 28.5$ deg. The perturbations are due to the principal gravitational harmonics and the solar/lunar gravitational accelerations. The satellite trace repetition parameter Q is the number of satellite revolutions that occur during one rotation of the Earth relative to the osculating orbit plane. Q is approximately the number of satellite revolutions per day.

11.4   Highly Eccentric, Critically Inclined $Q = 2$ Orbits (Molniya)

This discussion is devoted to examining the characteristics of a very specialized orbit: the $Q = 2$ orbit, which is highly eccentric and critically inclined, with apogee located over the northern hemisphere. This type of orbit has the useful characteristic of enabling observation of vast areas of the northern hemisphere for extended periods of time each day. Typically, two properly phased spacecraft located in two ideal ground track locations (groundtracks repeat day after day) will view continuously 55–60% of the northern hemisphere, centered at the North Pole, as illustrated in Fig. 11.20.

Orbital Geometry

The following paragraphs examine each of the classical orbital elements, and optimal values are assigned to those elements for which such assignment has meaning.

Semimajor Axis

The semimajor axis a is determined by calculation to be approximately 26,554 km. This value assures that the groundtrack of the orbit will remain fixed relative

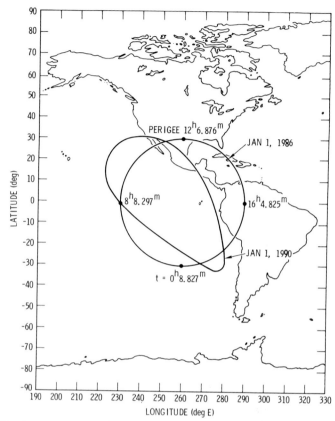

Fig. 11.19   Four-year ground trace change due to Earth's gravitational harmonic effects.

to the Earth and will repeat itself every day ($Q = 2$, an integer, assures this). Operationally, there are forces that tend to alter a (drag, solar pressure, tesseral harmonics, etc.), and fuel must be expended by the spacecraft to make periodic corrections to a.

When $a \approx 26,554$ km, the Keplerian period is 11.967 mean solar hours, which is one-half of the quantity 360 deg (one complete Earth rotation on its axis relative to the stars) divided by 15.041067 deg/h (the Earth's rotation rate). And so, to be a repeating groundtrack, $Q = 2$ orbit, the period is not 12 h (or one-half a mean solar day), but 11.967 h, which is one-half of a mean sidereal day.

Eccentricity

The value of eccentricity e will vary over a typical mission lifetime, largely because of solar and lunar gravitational perturbations. Eccentricities ranging from 0.69 to 0.74 are typical of this orbit type. The initial mission value of e is dictated by launch date, mission duration, minimum acceptable value of h_p, groundtrack location, and initial right ascension of ascending node (Ω). An adequately accurate computer program must be utilized to assure that the spacecraft will not re-enter the atmosphere at a premature time.

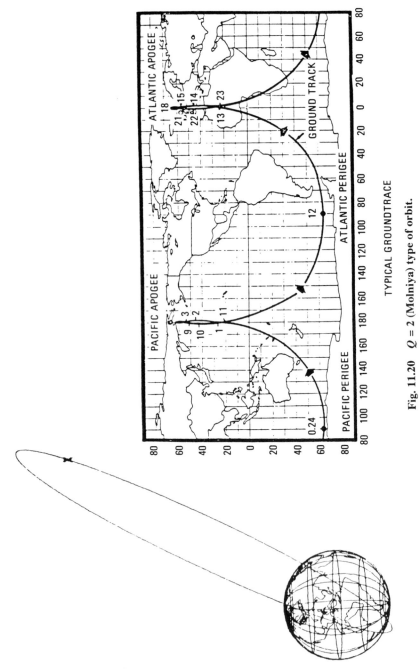

Fig. 11.20 $Q = 2$ (Molniya) type of orbit.

Fig. 11.20 $Q = 2$ (Molniya) type of orbit.

Inclination

For observation of northern hemisphere regions, the inclination should be high. In order that apogee be maintained at the northernmost point for long periods of time, the inclination must be approximately 63.44 deg (the critical inclination). Inclinations higher than 63.44 deg will force the line of apsides to rotate in a direction opposite the direction of satellite motion. Inclinations lower than 63.44 deg will rotate the line of apsides in the same direction as the satellite motion. These statements are valid for direct orbits only. The arguments are reversed for retrograde motion.

Right Ascension of Ascending Node

The node of the orbit can take any value, 0–360 deg. The value chosen essentially pinpoints the time of day of launch. At launch, the node (or time of day) must be carefully chosen so that lunisolar perturbations do not act to reduce h_p below an acceptable level during the nominal mission duration.

Argument of Perigee

In order that optimal visibility of the northern hemisphere be maintained, ω should be held as close to 270 deg as possible. Selecting $i = 63.44$ deg assures that ω will drive away from 270 deg by less than ± 5 deg for mission lifetimes on the order of 5 yr.

11.5 Frozen Orbits

Introduction

The term *frozen orbit* first appeared in the literature in 1978 in a paper by Cutting et al. entitled "Orbit Analysis for SEASAT-A."[3] A frozen orbit is one whose mean elements, specifically, eccentricity e and argument of perigee ω, have been selected to produce constant values, or nearly so, of e and ω with time. Thus, perigee rotation is stopped, and the argument of perigee is frozen at 90 deg. Frozen orbits maintain a constant altitude profile over the oblate Earth from revolution to revolution, with very small altitude variations over the northern hemisphere. This profile appears to decay uniformly because of atmospheric drag. Thus, frozen orbits are particularly useful for low-altitude missions that require eccentricity control. If radiation pressure and, perhaps, drag are not too influential, frozen orbit eccentricity will remain nearly constant for years. Otherwise, active maneuvering to maintain the frozen orbit is feasible and has been demonstrated (see Ref. 4). Frozen orbits are necessary for missions that require tight longitudinal control of the ground trace because there are no longitudinal variations of the ground trace due to perigee rotation.

The orbit will remain frozen in time if only zonal harmonics are assumed in the geopotential, i.e., no tesseral harmonics, and if atmospheric drag, solar-radiation pressure, and third-body gravitational perturbations are ignored.

Geometric interpretations of the first two zonal harmonics, J_2 and J_3, are shown in Fig. 11.21. The Earth's oblateness is caused by a combination of J_2 and the rotation of the Earth. The difference between the polar and equatorial radii of the Earth is 21.4 km. The pear shape of the Earth is caused by J_3. The height of the

SPHERICAL EARTH + ROTATION + J$_2$

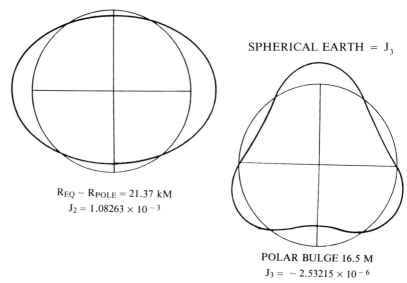

SPHERICAL EARTH = J$_3$

R$_{EQ}$ − R$_{POLE}$ = 21.37 kM
J$_2$ = 1.08263 × 10^{-3}

POLAR BULGE 16.5 M
J$_3$ = − 2.53215 × 10^{-6}

Fig. 11.21   Shape of the Earth.

bulge at the North Pole is about 16.5 m. For the Earth, the value of J_3 is three orders of magnitude less than the value of J_2.

Variational Rate Equations

Equations (11.20) and (11.21) are the averaged variational rate equations for eccentricity and argument of perigee for an Earth model consisting of a rotating Earth plus zonal harmonics J_2 and J_3 (see Chapter 9). All elements are mean elements:

$$\frac{de}{dt} = \frac{3J_3 n}{2(1 - e^2)^2} \left(\frac{R}{a}\right)^3 \sin i \left(1 - \frac{5}{4} \sin^2 i\right) \cos \omega \qquad (11.20)$$

$$\frac{d\omega}{dt} = \frac{3J_2 n}{(1 - e^2)^2} \left(\frac{R}{a}\right)^2 \left(1 - \frac{5}{4} \sin^2 i\right) \left[1 + \frac{J_3}{2J_2(1 - e^2)}\right. \qquad (11.21)$$
$$\left. \times \left(\frac{R}{a}\right) \frac{\sin i \sin \omega}{e}\right]$$

where

n = the satellite mean motion
R = the Earth's mean equatorial radius

Mean elements reflect secular and long-period perturbative variations. Mean elements can be obtained from osculating elements by averaging out the short-period oscillations.

Note that there is no J_2 term in Eq. (11.20) but that Eq. (11.21) contains both J_2 and J_3. Also, note that both equations contain the well-known expression $(1 - 5/4\sin^2 i)$. These equations equal zero when this expression equals zero, i.e., when $i = 63.4$ deg or 116.6 deg, the critical inclination. These are frozen-orbit solutions. This solution was probably first implemented in 1964 with the advent of the Molniya satellite orbit.

However, the frozen-orbit solutions of interest are implemented by setting $\omega = 90$ deg, so that $de/dt = 0$, and by setting the bracketed term in Eq. (11.21) equal to zero, so that $d\omega/dt = 0$. From this term, the frozen mean eccentricity can be solved for given values of the mean semimajor axis and mean inclination. Note that the eccentricity [obtained as a solution to Eq. (11.22)] is

$$e \approx -\frac{J_3}{2J_2}\left(\frac{R}{a}\right)\sin i \qquad (11.22)$$

It is of the order of 10^{-3} because J_3 is three orders of magnitude less than J_2 (see Fig. 11.21).

Frozen-Orbit Solutions

Results from Eq. (11.22) are presented in Fig. 11.22 as the dashed line. Extensive numerical investigations of frozen-eccentricity solutions have revealed the importance of higher-order zonal harmonics, especially for inclinations near the critical inclination. The solid line in Fig. 11.22 describes frozen-eccentricity solutions obtained by using the $J_2–J_{12}$ zonal harmonic terms in the WGS-84 geopotential model.

There are distinct behaviors of these solutions in three regions of inclination: $i < 63.4$ deg, 63.4 deg $< i <$ 66.9 deg, and 66.9 deg $< i \le$ 90 deg. At $i = 10$ deg, the frozen eccentricity is 0.132×10^{-3}. As inclination increases, the frozen eccentricity increases in an oscillatory manner as shown in Fig. 11.22. The frozen-eccentricity increases only slightly between $i = 35$ deg and $i = 40$ deg. The frozen-eccentricity increases dramatically, however, as the inclination approaches the critical value, i.e., $i = 63.435$ deg. A frozen solution, $e = 10.1 \times 10^{-3}$, was found at $i = 63.0$ deg. Presumably, frozen orbits with larger eccentricities can be found in the inclination region between 63.0 and 63.435 deg.

Inverted frozen orbits were discovered for inclinations slightly greater than 63.435 deg. Inverted frozen-orbit solutions occur at $\omega = 270$ deg rather than $\omega = 90$ deg. Thus, perigee is at the southernmost point in the orbit rather than at the northernmost point. The initial discovery of inverted frozen orbits was published by J. C. Smith in 1986.[6] At $i = 63.6$ deg, the frozen eccentricity was found to be 23.5×10^{-3}, or 0.0235. As inclination increases, the frozen eccentricity decreases dramatically. At $i = 66.6$ deg, the frozen solution is $e = 0.124 \times 10^{-3}$ and, at $i = 66.8$ deg, $e = 0.041 \times 10^{-3}$.

Then, as inclination increases slightly, a transition in frozen solutions from $\omega = 270$ deg to $\omega = 90$ deg occurs, so that, for $i = 67.1$ deg, $e = 0.068 \times 10^{-3}$ at $\omega = 90$ deg. And, for a narrow range of inclinations, no frozen-orbit solutions could be found. As inclination increases from $i = 67.1$ deg, eccentricity increases rapidly to $e = 0.65 \times 10^{-3}$ at $i = 70.0$ deg. Then, as Fig. 11.22 shows, the frozen eccentricity increases less rapidly with increasing inclination until $e = 1.26 \times$

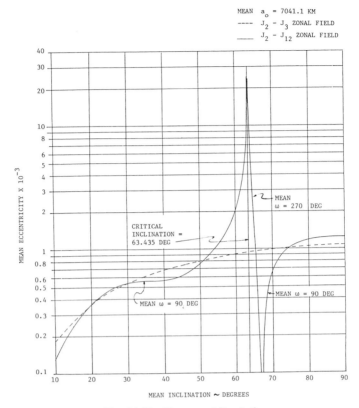

Fig. 11.22   Frozen-orbit solutions.

10^{-3} at $i = 90.0$ deg. Thus, the curve corresponding to the $J_2 - J_{12}$ zonal field is quite different from the curve corresponding to the J_2 and J_3 zonal field, especially for inclinations between 50 and 75 deg. In the vicinity of the critical inclination, the values of de/dt and $d\omega/dt$ from Eqs. (11.20) and (11.21) are near zero. Terms involving higher zonal harmonics than J_2 and J_3 then become relatively more important.

Circulations About the Frozen-Orbit Conditions

It is useful and interesting to examine the behavior of mean e and mean ω for initial values that are near the frozen-orbit values. For nearby initial values, e and ω will circulate about the frozen-orbit conditions as illustrated in Fig. 11.23. This figure is from Ref. 5. For $i = 98.7$ deg and a semimajor axis value of 7198.7 km, the frozen eccentricity is 1.15×10^{-3} at $\omega = 90$ deg. Although the semimajor axis is slightly different, by about 160 km, the value of frozen e is very nearly the value obtained from Fig. 11.22 at an inclination of 81.3 deg, i.e., the supplement of 98.7 deg.

For initial conditions that are near, but not at, the frozen point, e and ω will move counterclockwise in closed contours. For inclinations less than 63.435 deg

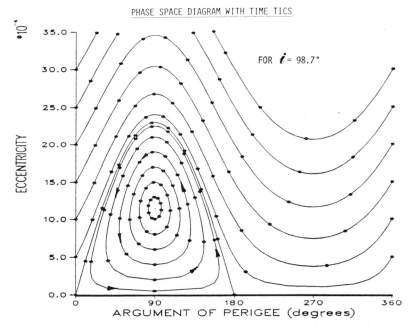

Fig. 11.23 Phase space diagram marked at 13-day interval (from Ref. 5).

or greater than 116.565 deg, the motion is clockwise. The period of circulation is 360 deg/$\dot{\omega}$, where $\dot{\omega}$ is the familiar J_2 secular apsidal rate. A $J_3 - J_{21}$ zonal field was used to generate the circulation curves, and 13-day intervals in time were marked on the curves. The distribution of time tics on the larger closed curves is interesting. More time tics are clustered at the higher than the lower eccentricities. Thus, during one circulation, the eccentricity is larger than the frozen value for a longer period of time than it is smaller than the frozen value. For larger deviations from the frozen point, the contours do not close.

In Ref. 7, Rosborough and Ocampo derive and present frozen-orbit eccentricity for an Earth gravity field complete to degree 50. Their results show that inverted frozen orbits, i.e., $\omega = 270$ deg, occur in a low-altitude ($a < 1.09$ Earth radii), low-inclination ($i < 10$ deg) region as well as in a region of inclination 63.4 deg $< i <$ 66.9 deg.

References

[1]Ginsberg, L. J., and Luders, R. D., *Orbit Planner's Handbook,* The Aerospace Corp., El Segundo, CA, Technical Memorandum, 1976.

[2]Milliken, R. J., and Zoller, C. J., "Principle of Operation of Navstar and System Characteristics," *Navigation: Journal of the Institute of Navigation,* Vol. 25, Summer, 1978, pp. 95–106.

[3]Cutting, E., Born, G. H., and Frautnick, J. C., "Orbit Analysis for SEASAT-A," *Journal of the Astronautical Sciences,* Vol. XXVI, Oct.–Dec. 1978, pp. 315–342.

[4]McClain, W. D., "Eccentricity Control and the Frozen Orbit Concept for the Navy Remote Ocean Sensing System (NROSS) Mission," AAS Paper 87-516, Aug. 1987.

[5]Nickerson, K. G. et al., "Application of Altitude Control Techniques for Low Altitude Earth Satellites," *Journal of the Astronautical Sciences,* Vol. XXVI, April–June 1978, pp. 129–148.

[6]Smith, J. C., "Analysis and Application of Frozen Orbits for the Topex Mission," AIAA Paper 86-2069-CP, Aug. 1986.

[7]Rosborough, G. W., and Ocampo, C. A., "Influence of Higher Degree Zonals on the Frozen Orbit Eccentricity," AAS Paper 91-428, Aug. 1991.

[8]Milstead, A. H., "Launch Windows for Orbital Missions," The Aerospace Corp., El Segundo, CA, Report No. TDR-269(4550-10)-6, April 1, 1964.

12
Lunar and Interplanetary Trajectories

12.1 Introduction

The solar system consists of a single star (the sun) and nine principal planets that move around the sun in orbital paths that are nearly circular except for that of Pluto, which is highly eccentric. Early man considered the Earth to be the center of the universe and the five planets, Venus, Mercury, Mars, Jupiter, and Saturn, to be divine. The most ancient observations of the planets date back 2000 years B.C. and appear to come from the Babylonian and Minoan civilizations. The term *planet* means "wanderer" in Greek and refers to the celestial objects that move relative to the stars.

The Egyptians, Greeks, and Chinese once thought of Venus, for example, as two stars because it was visible first in the morning and then in the evening sky. The Babylonians called Venus "Istar," the personification of woman and the mother of the gods. In Egypt, the evening star was known as Quaiti and the morning star Tioumoutiri; to the Chinese, Venus was known as Tai-pe, or the Beautiful White One. The Greeks called the morning star Phosphorus and the evening star Hesperos but, by 500 B.C., the Greek philosopher Pythagoras had come to realize that the two were identical. As time passed, the Romans changed the name of the planet to honor their own goddess of love, Venus.

It was not until the Golden Age of Greece that astronomy as a science was placed on a firm foundation and the Earth and the planets were regarded as globes rather than flat surfaces. Had Greek quantitative analysis taken one more step and dethroned the world from its position as the center of the universe, the progress of human thought and logic would have been accelerated. The Greek philosopher and mathematician Aristarchus held a heliocentric view of the solar system, but his ideas were opposed on religious grounds, and the later Greeks reverted to the idea of a central Earth.

Ptolemy, who died in about A.D. 150, left a record of the state of the universe at the end of the classic Greek period. In his Ptolemaic system, the Earth lies in the center of the universe, with the various heavenly bodies revolving around it in perfect circles. First comes the moon, the closest body in the sky; then come Mercury, Venus, and the sun, followed by the three other planets then known (Mars, Jupiter, and Saturn), and finally the stars.[1]

An interesting law of planetary spacing was suggested by Bode in 1771, who stated that the normalized mean distances of the planets from the sun are given by the terms of the series 0, 3, 6, 12, etc., when added to 4 and divided by 10. This law implied that there should also be a planet between Mars and Jupiter, which led to the discovery of the asteroid Ceres and the asteroid belt. Bode's law holds for the planets closest to the sun [e.g., for Earth $(6 + 4)/10 = 1$ A.U. (astronomical unit, mean radius of the Earth orbit)] but fails for Neptune, where it predicts 38.3 vs 30.2 actual distance. It also appears to be valid for the satellites of the planets as, for example, those of Uranus. The planets and their relative size are illustrated in Fig. 12.1, and the planetary and satellite data are given in Tables 12.1 and 12.2, respectively.

This picture represents the relative sizes of the planets in the Solar System, compared with the Sun's disk. The distances from the Sun are given in millions of miles. As a visual reference, the Earth is about 110 Sun diameters away or about 11,700 times Earth's own diameter away. On the scale of this picture, Earth would actually be about 20 meters (65 ft) away from the Sun. Jupiter is 5 times as far, and Saturn nearly 10 times as far. Since Pluto's orbit is highly eccentric, it now is closer to the Sun than Neptune, which would be 30 times farther away than the Earth.

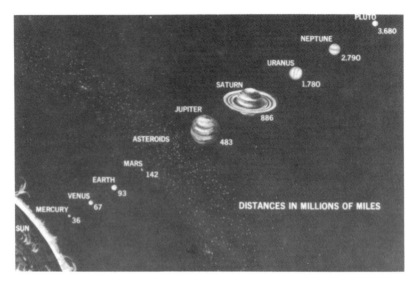

Fig. 12.1   The planets' relative sizes and mean distance from the sun.

Fundamental reasons for study of the solar system are exploration, questions related to the origin of the solar system, and asteroid mining missions, for example. Other potential applications include nuclear waste disposal missions and eventual colonization of the solar system. In this chapter, after a short historical background, the sphere of influence and the gravity-assist concepts are examined. Simple patched-conic approaches are then used to describe missions to the moon, and Mars.

12.2   Historical Background

During the 20-yr period from 1962 to 1981, the exploration of the solar system was initiated by means of unmanned spacecraft. Flybys of the inner planets Venus and Mars were accomplished (by Mariners 2 and 4–7 in the United States). In the 1970s, multiple flybys of the inner planets were performed, including orbiters about Mars and Venus (Mariner 9, Viking, and Pioneer Venus in the United States). Delivery of soft-landing vehicles to Mars and Venus was accomplished and flybys of Jupiter performed.

The 1980s began with the encounters of Pioneer 11 and two Voyager spacecraft with Saturn. During this period, the Soviet Union also sent a number of flyby, orbiting, atmospheric entry, and landing spacecraft to Venus and Mars. Complex terrain across several thousand miles of the surface of Venus was, for example, photographed by the Soviet Venera 16 imaging radar spacecraft, which revealed a

Table 12.1   Planetary data (from Ref. 8)

Planet	μ, km^3/s^2	Equatorial diameter, km	Mean distance from sun, 10^6 km	Sidereal period	Axial rotation (equatorial)	Axial inclination, deg	Mean Synodic period, days
Mercury	2.232×10^4	4670	57.9	88.0 days	58.7 days	?	115.9
Venus	3.257×10^5	12400	108	224.7 days	243 days	?	584.0
Earth	3.986×10^5	12700	149.6	365.3 days	23H 56M 04S	23° 27′	—
Mars	4.305×10^4	6760	227.7	687.0 days	24H 56M 23S	23° 59′	779.9
Jupiter	1.268×10^8	143000	777.8	11.86 yr	9H 50M 30S	3° 04′	398.9
Saturn	3.795×10^7	121000	1486	29.46 yr	10H 14M	26° 44′	378.1
Uranus	5.820×10^6	47100	2869	84.01 yr	10H 49M	97° 53′	369.7
Neptune	6.896×10^6	50700	4475	164.79 yr	About 14H	28° 48′	367.5
Pluto	3.587×10^5	5950	5899	248.43 yr	6D 9H	?	366.7

Planet	Escape velocity, km/s	Density: water = 1	Volume: Earth = 1	Mass: Earth = 1	Surface gravity: Earth = 1	Max surface, temperature, °F	Number of Satellites
Mercury	4.2	5.5	0.06	0.06	0.38	+770	0
Venus	10	5.3	0.86	0.82	0.90	+887	0
Earth	11	5.5	1	1	1	+140	1
Mars	6.4	3.9	0.15	0.11	0.38	+80	2
Jupiter	59.7	1.3	1319	318	2.64	-200	13
Saturn	35.4	0.7	744	95	1.16	-240	10
Uranus	22.4	1.7	47	15	1.11	-310	5
Neptune	31	1.8	54	17	1.21	-360	8
Pluto	?	?	0.1?	0.0026	?	?	1

Table 12.1   Satellite data (from Ref. 8)

Planet/Satellite	Mean distance from center of primary[a]	Sidereal period	Diam, km	Density: water = 1	Maximum magnitude	Reciprocal mass primary = 1	Discoverer
Earth							
Moon	385	27D 7H 43M	3470	3.3	12.7	81.3	—
Mars							
Phobos	9.33	7H 39M	23.3	?	11	?	Hall, 1877
Deimos	23.5	1D 6H 18M	11.3	?	12	?	Hall, 1877
Jupiter							
Amalthea (V)	182	11H 57M	241	?	13	?	Barnard, 1892
Io (I)	422	1D 18H 28M	3710	4.1	5.5	26,200	Galileo, 1609
Europa (II)	671	3D 13H 14M	3140	3.7	5.7	40,300	Galileo, 1609
Ganymede (III)	1,070	7D 3H 43M	5150	2.4	5.1	12,200	Galileo, 1609
Callisto (IV)	1,880	16D 16H 32M	4820	2.0	6.3	19,600	Galileo, 1609
Hestia (VI)	11,400	250D 26H	161	?	13.7	?	Perrine, 1904
Hera (VII)	11,700	259D 16H	56.3	?	17	?	Perrine, 1905
Demeter (X)	11,700	260D 12H	24.1	?	18.8	?	Nicolson, 1938
Adrastea (XII)	20,900	631D[b]	28	?	18.9	?	Nicolson, 1951
Pan (XI)	22,500	692D[b]	30.6	?	18.4	?	Nicolson, 1938
Poseidon (VIII)	23,500	744D[b]	56.3	?	18	?	Melotte, 1908
Hades (IX)	23,600	758D[b]	27.4	?	18.4	?	Nicolson, 1914
XIII	22,500	?[b]	8.04	?	?	?	Kowal, 1974

[a]Thousands of kilometers.
[b]Indicates retrograde motion.

Table 12.1 Satellite data (from Ref. 8) (cont.)

Planet/Satellite	Mean distance from center of primary[a]	Sidereal period	Diam, km	Density: water = 1	Maximum magnitude	Reciprocal mass primary = 1	Discoverer
Saturn							
Janus	158	17H 58M	241	?	14	?	Dollfus, 1966
Mimas	182	22H 37M	483	1	12	15,000,000	Herschel, 1789
Enceladus	240	1D 8H 53M	644	1	11	7,000,00	Herschel, 1789
Tethys	295	1D 21H 18M	1130	1.1	10.5	910,000	G. D. Cassini, 1684
Dione	378	2D 17H 41M	1450	3.2	10.5	910,000	G. D. Cassini, 1684
Rhea	528	4D 17H 25M	1770	2	9.3	250,000	G. D. Cassini, 1672
Titan	1,889	15D 22H 41M	5310	2.3	8.3	4,150	Huygens, 1655
Hyperion	1,488	21D 6H 38M	322	3	13	5,000,000	Bond, 1848
Lapetus	3,540	79D 7H 56M	2410	?	9	?	G. D. Cassini, 1671
Phoebe	13,000	550D[b] 10H 50M	241	?	14	?	Pickering, 1898
Uranus							
Miranda	122	1D 9H 50M	322	5	17	1,000,000	Kuiper, 1948
Ariel	192	2D 12H 29M	2410	5	14	67,000	Lassell, 1851
Umbriel	267	4D 3H 28M	1290	4	14.7	170,000	Lassell, 1851
Titania	438	8D 16H 56M	2410	6	14	20,000	Herschel, 1787
Oberon	586	13D 11H 7M	2410	5	14	34,000	Herschel, 1787
Neptune							
Triton	354	550D[b] 10H 50M	4830	?	14	750	Lassell, 1846
Nereid	5,630	550D[b] 10H 50M	322	?	14	?	Kuiper, 1949

[a]Thousands of kilometers.
[b]Indicates retrograde motion.

nearly 11-km-high mountain topped with a 100-km-diam crater, believed to be a massive meteorite impact crater.[2]

Preliminary explorations of comets and the continuing Voyager 2 mission to Uranus and Neptune were performed, with intensive investigation of the outer planets Saturn, Uranus, and Neptune, along with Saturn's satellite Titan.[3] Voyager 2's encounter with Uranus on 24 January 1986 showed, for example, that the planet's magnetic pole is inclined from the pole of rotation by an angle of about 55 deg, which is the largest inclination in the solar system. Images of Miranda and Umbriel, the satellites of Uranus, revealed three terrain types of different age and geology, such as hills, groved valleys, and craters.

One of the major exploration projects in the United States was the Galileo mission to Jupiter originally planned for 1986 but subsequently delayed to October 1989 because of the grounding of the Space Shuttle fleet and the new safety requirements posed by loss of Shuttle Centaur upper stage. A possible backup mission was also planned for July 1991 if the 1989 mission was not possible.

The trajectory to Jupiter employed a two-stage inertial upper stage (IUS) using one gravity-assist maneuver at Venus and two maneuvers at Earth, requiring more than 6 years of travel time.[4]

The new trajectory, which seemed to send Galileo on a cruise through the solar system, soon earned the name "Solar Cruiser."

GALILEO VEEGA TRAJECTORY TO JUPITER

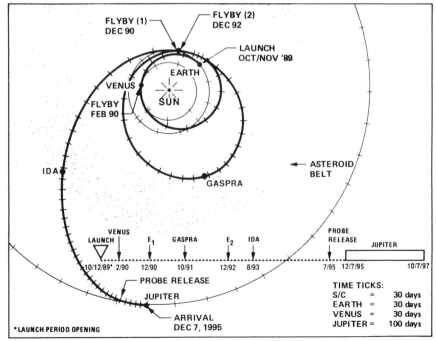

Fig. 12.2   Galileo's route to Jupiter.

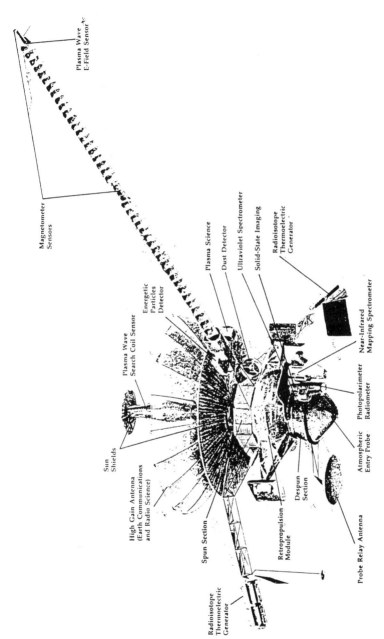

Fig. 12.3   Galileo's spacecraft.

Galileo was launched in October 1989 from Earth on a Space Shuttle and an inertial upper stage, a rocket whose energy is low compared to the Centaur. Instead of heading toward Jupiter or the asteroid belt, Galileo took a flight path that carried it to Venus. Galileo arrived there in February 1990. Venus' gravity accelerated Galileo and sent it on a flight path back toward Earth. When Galileo passed Earth in December 1990, the Earth's gravitational field added energy to send Galileo out to the asteroid belt. A propulsive maneuver, performed in December 1991, brought Galileo past Earth again in December 1992 for a last gravity assist before the spacecraft began its final path to Jupiter. Arrival at Jupiter occurred late in 1995.

A view of the Galileo trajectory and its spacecraft are shown in Figs. 12.2 and 12.3, respectively.

About 150 days before arrival of the Galileo spacecraft, the atmospheric probe separated from the Orbiter. The Orbiter then flew within 1000 km of the satellite Io, whose gravitational field helped to slow the spacecraft. The probe is designed to sample Jupiter's equatorial zone, which consists primarily of ammonia.

According to the plan, the orbiter entered a Jovian orbit ranging from 200,000 km to more than 10 million km for a 20-month study of Jupiter's environment and its satellites. Current knowledge of this environment is illustrated in Figs. 12.4–12.6.

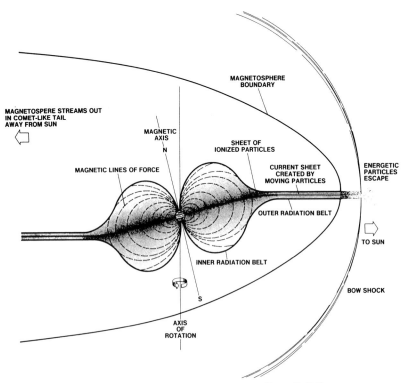

Fig. 12.4   Jupiter's magnetosphere (from Ref. 8).

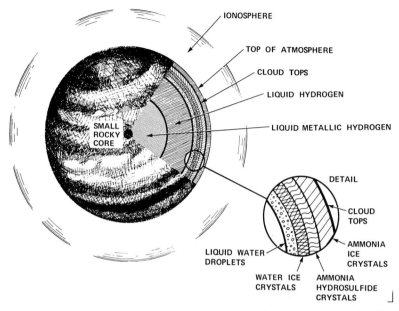

Fig. 12.5   Jupiter atmosphere model (scale exaggerated); atmosphere depth to liquid zone is 1000 km (600 miles) (from Ref. 8).

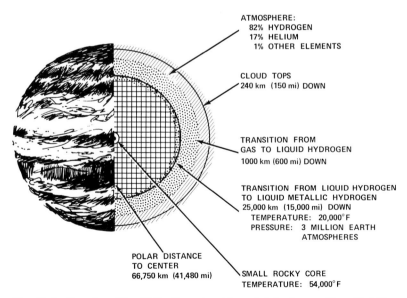

Fig. 12.6   Interior of liquid Jupiter; planet is mainly hydrogen (from Ref. 8).

12.3 Important Concepts

Sphere of Gravitation[5]

Consider the motion of a mass point (P, m) under the influence of two larger masses m_1 and m_2, as shown in Fig. 12.7. The larger masses are termed centers of attraction A_1 and A_2. It is assumed that $m_1 \ll m_2$ and $m \ll m_1$.

The values of the gravitational forces F_1 and F_2 acting on mass m toward A_1 and A_2 are with reference to Fig. 12.7 given by

$$F_1 = \frac{Gmm_1}{|A_1 P|^2}, \qquad F_2 = \frac{Gmm_2}{|A_2 P|^2} \qquad (12.1)$$

where G = universal constant of gravitation.

The locus of points where $F_1 > F_2$ defines the sphere of gravitational attraction of mass m_1 with respect to mass m_2. The location and the radius of the sphere are determined from boundary condition

$$F_1 = F_2$$

or

$$\frac{A_1 P}{A_2 P} = \sqrt{\frac{m_1}{m_2}} = \text{const} < 1 \qquad (12.2)$$

which indicates that the ratio of the distance from P to A_1 and A_2 is constant. From elementary geometry, the locus of points defined by this condition is a sphere, the diameter of which is defined by points C and D in Fig. 12.7.

Thus, letting R_s, R_o be the radius and origin of the sphere of gravitation and using Eq. 12.1 for the collinear points of attractions C and D, the ratio of the distances

$$\frac{A_1 C}{A_2 C} = \frac{A_1 D}{A_2 D} = \sqrt{\frac{m_1}{m_2}} \qquad (12.3)$$

or

$$\frac{R_s - R_o}{A_1 A_2 - (R_s - R_o)} = \frac{R_s + R_o}{A_1 A_2 + (R_s + R_o)} = \sqrt{\frac{m_1}{m_2}} \qquad (12.4)$$

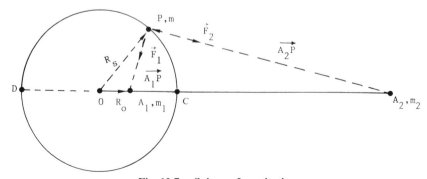

Fig. 12.7   Sphere of gravitation.

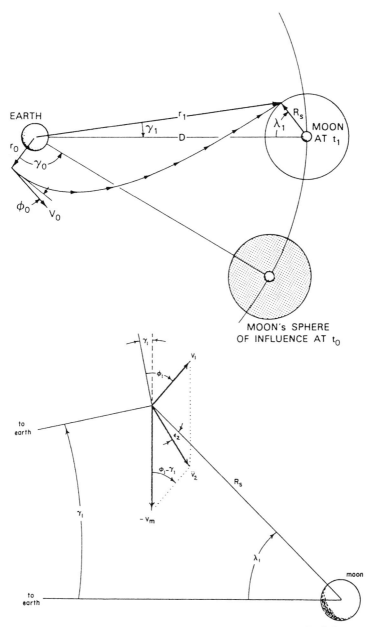

Fig. 12.15   Earth-moon patched-conic geometry (from Ref. 6).

$$\phi_1 = \cos^{-1}\left(\frac{h}{r_1 v_1}\right) \tag{12.28}$$

$$\gamma_1 = \sin^{-1}\left(\frac{R_s}{r_1}\sin\lambda_1\right) \tag{12.29}$$

For the velocity of the moon,

$$v_m = 1.018 \text{ km/s}$$

$$v_2 = \sqrt{v_1^2 + v_m^2 - 2v_1 v_m \cos(\phi_1 - \gamma_1)} \tag{12.30}$$

$$E_m = \frac{v_2^2}{2} - \frac{\mu_m}{R_s} \tag{12.31}$$

where $\mu_m = 4.093 \times 10^3 \text{ km}^3/\text{s}^2$

$$h_m = R_s v_2 \sin\varepsilon_2 \tag{12.32}$$

where

$$\varepsilon_2 = \sin^{-1}\left[\frac{v_m}{v_2}\cos\lambda_1 - \frac{v_1}{v_2}\cos(\lambda_1 - \gamma_1 - \phi_1)\right] \tag{12.33}$$

$$P_m = \frac{h_m^2}{\mu_m} \tag{12.34}$$

$$e_m = \sqrt{1 + 2\frac{E_m h_m^2}{\mu_m^2}} \tag{12.35}$$

$$r_p = \frac{P_m}{1 + e_m} \tag{12.36}$$

$$v_p = \sqrt{2\left(E_m + \frac{\mu_m}{r_p}\right)} \tag{12.37}$$

Here,

$$p = \frac{h^2}{\mu_e} \tag{12.38}$$

$$a = \frac{-\mu_e}{2E_e} \tag{12.39}$$

$$e = \sqrt{1 - \frac{p}{a}} \tag{12.40}$$

For true anomalies, $\theta_0, \theta_1,$

$$\cos\theta_0 = \frac{p - r_0}{r_0 e} \rightarrow \theta_0 \tag{12.41}$$

$$\cos \theta_1 = \frac{p - r_1}{r_1 e} \to \theta_1 \qquad (12.42)$$

$$E_0 = \cos^{-1} \left(\frac{e + \cos \theta_0}{1 + e \cos \theta_0} \right) \qquad (12.43)$$

$$E_1 = \cos^{-1} \left(\frac{e + \cos \theta_1}{1 + e \cos \theta_1} \right) \qquad (12.44)$$

Flight time Δt is given by

$$\Delta t = t_1 - t_0 = \sqrt{\frac{a^3}{\mu_e}} [(E_1 - e \sin E_1) - (E_0 - e \sin E_0)] \qquad (12.45)$$

Phase angle at departure is found from

$$\gamma_0 = \theta_1 - \theta_0 - \gamma_1 - \omega_m \Delta t \qquad (12.46)$$

where $\omega_m = 2.649 \times 10^{-6}$ rad/s.

Typical flight time as a function of injection velocity at 320 km altitude for the phase angle at arrival $\lambda_1 = 65$ deg is shown in Fig. 12.16. The corresponding perigee radius and velocity at the moon (perilune) are shown in Figs. 12.17 and 12.18, respectively.

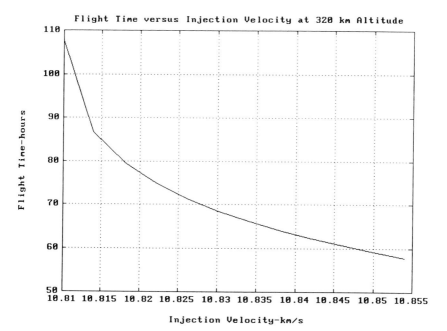

Fig. 12.16   Flight time vs injection velocity at $R_0 = 320$ km.

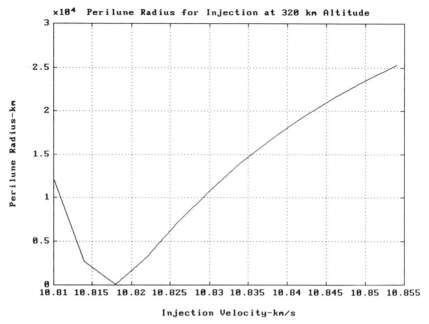

Fig. 12.17   Perilune radius for injection at 320 km altitude.

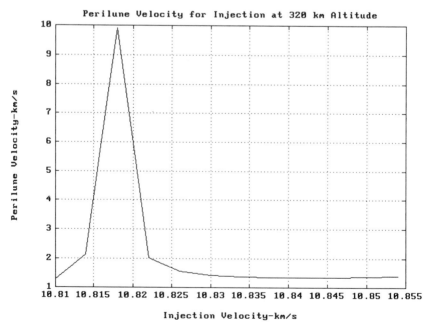

Fig. 12.18   Perilune velocity for injection at 320 km altitude.

12.6　Three-Dimensional Trajectories

The three-dimensional trajectories require the specification of several parameters in addition to those for the two-dimensional case discussed previously. The additional parameters are the initial latitude β, the initial launch azimuth ψ, the longitude difference between the initial point and the moon at impact $\Delta\lambda$, the instantaneous declination of the moon δ, and its maximum value δ_m. The relationships among these parameters determine the launch window at any launch site and the conditions at arrival at the moon. Reference [10] shows, for example, that these families of Earth-moon trajectories departing in either co- or counter-direction with Earth rotation have a common vertex on the far side of the moon (see Fig. 12.19). The corotational direction takes advantage of Earth tangential velocity at the launch site. An example of a Ranger-type (lunar impact) trajectory is illustrated in Fig. 12.20.

The targeting parameter B is defined as a vector originating at the center of the target body (e.g., moon) and directed perpendicular to the incoming asymptote of the target-centered approach hyperbola. The targeting parameter B is resolved into two components that lie in a plane normal to the incoming asymptote Si. The orientation of the reference axes in this plane is arbitrary but is usually selected to lie in a fixed plane. For interplanetary trajectories, the unit vector $\hat{T}$ is in the ecliptic plane, and $\hat{R}$ is normal to it, as shown in Fig. 12.21.

12.7　Interplanetary Trajectories

Types of Transfers

Feasibility-type trajectories that are used for preliminary vehicle performance studies involve a simple two-body problem. The classical Hohmann trajectories, for example, are ellipses that are tangent to both the launch and the arrival orbit. The energy for transfer from the launch orbit to the target orbit is, in most cases, a minimum for this trajectory, but the transfer time is usually quite long. The time can be found from Kepler's third law.

There are three main groups of trajectories that can be evolved from the Hohmann transfer: 1) staying tangential to the larger orbit but intersecting the smaller one, 2) intersecting the larger orbit and staying tangential to the smaller one, and 3) intersecting both orbits. Parabolic or hyperbolic transfer trajectories intersect the larger orbit and may, but do not need to be, tangential to the smaller orbit. The flight times for parabolic or hyperbolic trajectories are generally shorter than those for Hohmann transfers. Examples of several types of Earth-Mars trajectories are illustrated in Fig. 12.22.

Mars Landing Mission Example

Taking the approach of Ref. 6, we will examine the example of a spacecraft lander (probe) on Mars. For this case, the spacecraft is subject to the gravitational actions of Earth, sun, and Mars, as shown in Fig. 12.23.

The probe of mass m_p is subject to the gravitational forces F_s, F_e, F_m of sun, earth, and Mars, respectively; thus, the equation of motion is

$$m_p\ddot{r}_p = F_s + F_e + F_m \tag{12.47}$$

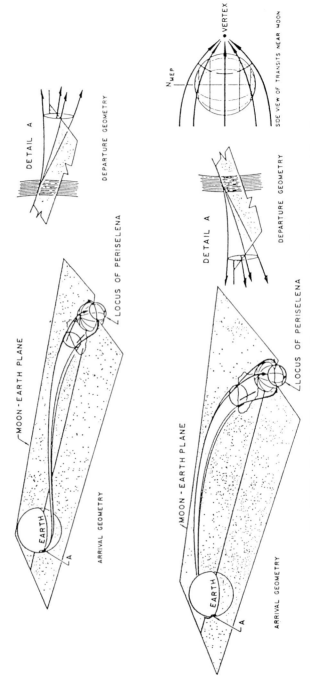

Fig. 12.19 Family of transits departing from moon-Earth plane equatorial perigees in counter-rotational directions (from Ref. 10).

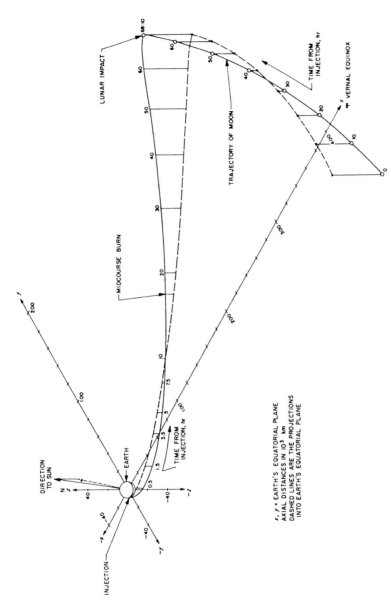

Fig. 12.20   Typical geocentric spatial trajectory trace (Ranger 11).

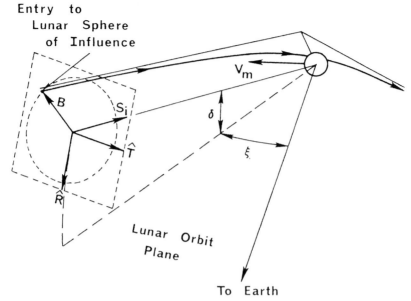

Entry to
Lunar   Sphere
of Influence

Fig. 12.21   Lunar targeting parameters.

or

$$m_p \ddot{\boldsymbol{r}}_p = \frac{-\mu_s}{r_p^3}\boldsymbol{r}_p - \frac{\mu_e(\boldsymbol{r}_p - \boldsymbol{r}_e)}{|\boldsymbol{r}_p - \boldsymbol{r}_e|^3} - \frac{\mu_m(\boldsymbol{r}_p - \boldsymbol{r}_m)}{|\boldsymbol{r}_p - \boldsymbol{r}_m|^3}$$

(12.48)

Here s, e, and m refer to the sun, Earth, and Mars, respectively.

This equation is not integrable in closed form and must be evaluated numerically. Approximate solutions for the trajectory can be obtained using a patched-conic approach. In this approach, the trajectory is divided into three different phases: 1) heliocentric, 2) Earth departure, and 3) Mars arrival. Each phase is a two-body Keplerian orbit, the conic section of which is "patched" with the following phase. Approximate velocity requirements for the mission can in this way be estimated to determine the feasibility of the mission.

Heliocentric phase.   Assuming a Hohmann-transfer heliocentric trajectory from Earth to Mars, as illustrated in Fig. 12.24, the required perigee velocity with respect to the sun is given by the equation

$$v_{\text{pt}} = \sqrt{\mu_s\left(\frac{2}{r_e} - \frac{1}{a_t}\right)}$$

(12.49)

$$= 32.74 \text{ km/s}$$

where

$r_e = 1 \text{ A.U.} = 1.49597893 \times 10^8 \text{ km}$

$a_t = \dfrac{r_e + r_m}{2} = \dfrac{2.523}{2}$

$\quad = 1.262 \text{ A.U.}$

$\quad = 1.887 \times 10^8 \text{ km}$

$\mu_s = 1.32712499 \times 10^{11} \text{ km}^3/\text{s}^2$

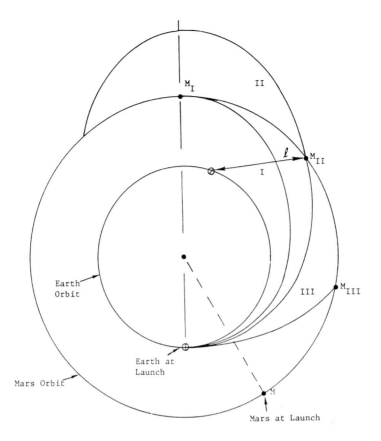

Trajectory	V_{bo}, km/s	V_∞, km/s	Flight time, days	Arrival Earth-Mars distance l, 10^6 km
I Hohmann	11.6	2.94	259	236
II Elliptic	11.9	3.75	165	133
III Parabolic	16.7	12.33	70	79

V_{bo} = burnout velocity at 500 km (v_c = 7.62 km/s)
V_∞ = velocity at infinity WRT Earth

Fig. 12.22   Earth-Mars trajectory velocity requirements.

The transfer-orbit apogee velocity at Mars is

$$v_{at} = \sqrt{\mu_s\left(\frac{2}{r_m} - \frac{1}{a_t}\right)}$$

$$= 21.45 \text{ km/s}$$

(12.50)

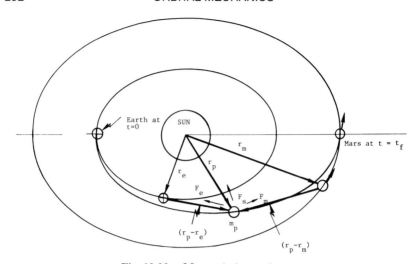

Fig. 12.23   Mars mission trajectory.

The velocity at exit from the Earth sphere of influence is

$$v_\infty = v_{pt} - v_e$$

$$= 32.74 - 29.78 \qquad\qquad (12.51)$$

$$= 2.96 \text{ km/s}$$

Therefore, $C_3 = v_\infty^2 = 8.76 \text{ km}^2/\text{s}^2$, and the energy $E_e = C_3/2 = 4.38 \text{ km}^2/\text{s}^2$.

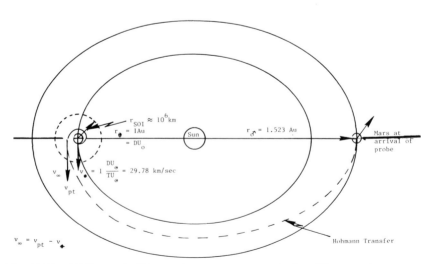

Fig. 12.24   Heliocentric Mars trajectory: $r_{SOI} \approx \ell \, (m_1/m_\odot)^{2/5}$; $\ell = a$ **for planet;** m_1 = **mass of planet;** $m_\odot$ **mass of sun.**

Earth departure phase.   Assuming an injection burn at an altitude of 300 km, the burnout velocity required is

$$v_{\text{bo}} = \sqrt{2\left[\frac{\mu_e}{(R_e + h)} + E_e\right]}$$ (12.52)

$$= 11.32 \text{ km/s}$$

where

$R_e = 6378$ km
    $=$ equatorial radius of Earth
$h\ \ = 300$ km
    $=$ altitude at injection
$\mu_e = 3.9860064 \times 10^5 \text{ km}^3/\text{s}^2$
    $=$ gravitational parameter of earth
$E_e = 4.38 \text{ km}^2/\text{s}^2$
    $=$ energy of escape hyperbola

Mars arrival phase.   For a soft landing at Mars, the energy of the hyperbolic orbit at Mars is

$$E_m = \frac{v_\infty^2}{2}$$

$$= \frac{(v_m - v_{at})^2}{2}$$

$$= \frac{(24.13 - 21.45)^2}{2}$$ (12.53)

$$= 3.59 \text{ km}^2/\text{s}^2$$

where $v_m = 24.13$ km/s $=$ velocity of Mars. Retro velocity at Mars surface is

$$v_{\text{retro}} = \sqrt{2\left(\frac{\mu_m}{R_m} + E_m\right)}$$

$$= 5.71 \text{ km/s}$$

where

$\mu_m = 43058 \text{ km}^3/\text{s}^2$
    $=$ gravitational parameter of Mars
$R_m = 3379$ km
    $=$ equatorial radius of Mars
$E_m = 3.59 \text{ km}^2/\text{s}^2$

At Mars SOI

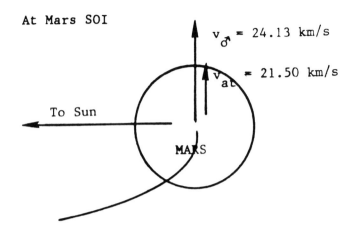

$v_{\sigma} = 24.13$ km/s

$v_{a} = 21.50$ km/s

To Sun

MARS

Total velocity requirements for the mission are

Burnout velocity at Earth $v_{bo} = 11.32$ km/s

Retro velocity at Mars $v_{retro} = \underline{5.71}$ km/s

Total   17.03 km/s

An equal amount of velocity impulse would be required for the return trip to Earth. Small thrust application may also be necessary for midcourse correction of the heliocentric trajectory, and the probe would retrofire first to deboost to a parking orbit before landing. The velocities that were computed are feasibility-type numbers. Precision velocities would be somewhat different as they must be obtained by integration of the equations of motion.

References

[1]Mariner-Venus 1962 Final Project Report, NASA SP-59, 1965.

[2]*Aviation Week and Space Technology*, Feb. 1985.

[3]Wood, L. J., and Jordan, J. F., "Interplanetary Navigation Through the Year 2005: The Inner Solar System," *Journal of the Astronomical Sciences*, Vol. 32, Oct.–Dec. 1984.

[4]*Aviation Week and Space Technology*, Dec. 1987.

[5]Balk, M. B., *Elements of Dynamics of Cosmic Flight, Nauka*, 1965 (in Russian).

[6]Bate, R. R., Mueller, D. D., and White, J. E., *Fundamentals of Astrodynamics*, Dover, New York, 1971.

[7]Clarke, T. C., and Fanale, F. P., "Galileo: The Earth Encounters," *Planetary Report*, Jet Propulsion Laboratory, CIT, Sept./Oct. 1989.

[8]"Pioneer to Jupiter," Bendix Field Engineering Corp., Palo Alto, CA, Nov. 1974.

[9]White, J. F., *Flight Performance Handbook for Powered Flight Operations*, Wiley, New York, 1963, pp. 2–54.

[10]Hoolker, R. F., and Braud, N. J., "Mapping the Course for the Moon Trip," *Astronautics & Aeronautics*, Feb. 1964.

Problems

12.1. Consider a nuclear waste disposal mission from Earth (1 A.U.) to a helio-centric circular orbit of radius 0.86 A.U.

a) What is the charcteristic energy C_3 required for a Hohmann transfer from Earth to the circular orbit or radius 0.86 A.U.?

b) What is the flight time via the Hohmann-transfer orbit from Earth to the 0.86 A.U. orbit about the sun?

12.2. To accomplish certain measurements of phenomena associated with sunspot activity, it is necessary to establish a heliocentric orbit with a perihelion of 0.85 A.U. The departure from the Earth's orbit will be at apohelion. What must the burnout velocity be at an altitude of 1300 km to accomplish this mission?

12.3. A Venus probe departs from a 6378-km altitude circular parking orbit with a burnout speed of 8.69 km/s. Find the hyperbolic excess speed at infinity.

12.4. Calculate the sphere of influence for the nine planets in the solar system.

12.5. Compute the distance of L_1, the Lagrangian liberation point, from the center of the moon along the Earth-moon line. To a first approximation, L_1 is an equilibrium point between the gravitational and centrifugal accelerations of the attracting bodies.

12.6. Compute the stationkeeping requirements (ΔV) to remain within 10 km of L_1 in Problem 12.5 for a year.

12.7. Compute the velocity impulse (ΔV) for transfer from a 100-km circular orbit at Mars to a hyperbolic orbit Earth return trajectory with an eccentricity $e = 1.5$.

12.8. Write the equation of motion for a solar sail in the solar system. Discuss the type of trajectories possible.

12.9. If the Earth were stopped in its orbit, what would be the elapsed time, in days, until collision with the sun? Assume point masses, and assume the Earth's orbit to be circular, with $r = 1.0$ A.U.

Selected Solutions

12.1. 1.28 km/s, 163.7 days

12.2. 10.26 km/s

12.3. 3.6 km/s

13
Space Debris

13.1 Introduction

Although the natural meteoroid environment has been considered in the designs of past and existing spacecraft, future satellite designs will have to take account of space debris in addition to the natural environment.

Man-made space debris differs from natural meteoroids because it is in permanent Earth orbit during its lifetime and is not transient through the regions of interest. As a consequence, a given mass of material presents a greater problem in the design and operation of spacecraft because of the extended time period over which there is risk of collision.

Past design practices and deliberate and inadvertent explosions in space have created a significant debris population in operationally important orbits. The debris consists of spent spacecraft and rocket stages, separation devices, and products of explosion. Much of this debris is resident at altitudes of considerable operational interest. Products larger than 10 cm^2 in low orbits can be observed directly. The existence of a substantially larger population of small fragments can be inferred from terrestrial tests in which the particle distributions from explosions have been assayed. From these tests it is reasonable to infer small particle numbers, of the order of 10,000 for each low-intensity explosion and several million for high-intensity explosions.

Two types of space debris are of concern: 1) large objects whose population, while small in absolute terms, is large relative to the population of similar masses in the natural flux (by a factor of about 1000); and 2) a large number of smaller objects whose size distribution approximates natural meteoroids. The interaction of these two classes of objects, combined with their long residual times in orbit, leads to the further concern that inevitably there will be collisions producing additional fragments and causing the total population to grow rapidly.

Some efforts to provide a definitive assessment of the orbiting debris problem have been and are being made by various government agencies and international organizations. Principal areas of concern are the hazards related to the tracked (cataloged), untracked, and future debris populations. Studies are being conducted in the areas of technology, space vehicle design, and operational procedures. Among these are ground- and space-based detection techniques, comprehensive models of Earth-space environment, spacecraft designs to limit accidental explosions, and different collision-hazard assessment methods. Occasional collision avoidance and orbit-transfer maneuvers are being implemented for selected satellites in geosynchronous orbits. The results and experience gained from the activities will, in time, create a better understanding of the problem and all its implications so that appropriate actions can be taken to maintain a relatively low-risk environment for future satellite systems.

A key aspect of the on-orbit debris hazard is that it is self-perpetuating. This arises from three factors: 1) A single spacecraft launch can be responsible for a multitude of hazardous objects in space; 2) orbital debris tends to disperse randomly, producing high intersection velocities and making avoidance extremely difficult; and 3) objects accumulate in Earth orbit rather than passing through the near-Earth space in the manner of meteoroids. Impact protection may not be feasible in most cases because of the likelihood of very high approach velocities and the fact that certain protuberances, especially those of relatively large areas such as solar arrays and antennas, cannot easily be shielded permanently. Evasive-maneuvering techniques may reduce the present probability of collision for specific satellites in certain circumstances but do not provide a practical long-term solution.

The only natural mechanism opposing debris buildup is removal by atmo-spheric drag. This process can take a very long time, however, especially from high altitudes, and causes debris to migrate from higher to lower altitudes. Another mechanism, collection by a spacecraft ("orbital garbage truck"), would be extremely difficult and expensive. Prevention of debris formation is the most effective approach.

At the present time, the collision hazard is real but not severe. Continuation of present policies and practices, however, ensures that the probability of collision will eventually reach unacceptable levels, perhaps within a decade. Future problems can be forestalled by initiating studies and implementing their results in five major areas: 1) education, 2) technology, 3) satellite and vehicle design, 4) operational procedures and practices, and 5) national and international space policies and treaties.

13.2   Space Debris Environment: Low Earth Orbit

At any one time, there are about 200 kg of meteoroid mass moving through altitudes below 2000 km at an average speed of about 20 km/s. Most of the mass is found in particles of about 0.1-mm diameter.[2] The meteoroid environment has always been a design consideration for spacecraft. The Apollo and Skylab spacecraft were built to withstand impacts on critical systems from meteoroids having sizes up to 3 mm in diameter. Larger sizes were so few in number as to be of no practical significance for the duration of the mission. Some small spacecraft systems required additional shielding against meteoroids as small as 0.3 mm in diameter in order to maintain an acceptable reliability. The trend in the design of future spacecraft (as for example, the Space Station) is toward larger structures, lighter construction, and longer times in orbit. These factors increase the concern about damage from particles in the 0.1- to 10-mm size range.

It is no longer sufficient, however, to consider only the natural meteoroid envi-ronment in spacecraft design. Since the time of the Apollo and Skylab programs, launch activity has continued and increased. As a result, the population of orbital debris has also increased substantially. The total mass of debris in orbit is now approximately 3 million kg at altitudes below 2000 km. Relative to one another, pieces of debris are moving at an average speed of 10 km/s, or only half the relative speed of meteoroids. The significant difference between the orbital de-bris population and the meteoroid population is that most of the debris mass is found in objects several meters in diameter rather than 0.1 mm in diameter as for meteoroids. This large reservoir of mass may be thought of as a potential source for particles in the 0.1- to 10-mm range. That is, if only one ten-thousandth of this mass were in this size range, the amount of debris would exceed the natural meteoroid environment. The potential sources for particles in this size range are many:

1) Explosions: More than 100 spacecraft are known to have exploded in low Earth orbit (LEO) and account for about 50% of the U.S. Space Command (USSPACECOM) Catalog. Since the fragment size distribution is a sensitive function of the intensity of the explosion, the number of smaller fragments produced by these explosions is not known.

2) Hypervelocity collision in space: One or two of the known satellite breakups may have been from hypervelocity collisions. The fragment size distribution of such a collision is known to include a large number of particles in the 0.1- to 10-mm range.

3) Deterioration of spacecraft surfaces: Oxygen erosion, ultraviolet radiation, and thermal stress are known to cause certain types of surfaces to deteriorate, producing small particles.

4) Solid rocket motor firings: One-third of the exhaust products of a solid rocket motor is aluminum oxide particles in the size range 0.0001 to 0.01 mm.

5) Unknown sources: Other sources are likely to exist. Particulates are commonly observed originating from the Shuttle and other objects in space.

13.3   Debris Measurements

What is currently known about the orbital debris flux is from a combination of ground-based and in-space measurements. These measurements have revealed an increasing population with decreasing size. Beginning with the largest sizes, a summary of these measurements follows.

USSPACECOM Catalog

The USSPACECOM is responsible for tracking and maintaining a catalog of "all man-made objects" in space. The catalog, as of April 1995, contained 7872 objects, most in LEO. Figure 13.1 shows the growth of the satellite population from 1959 through 1995.[3] This plot excludes space probes (over 100 of which were still in orbit). The linear growth rate of 240/yr provides a good approximation of the actual growth rate. The fluctuations in the curve are due primarily to satellite breakups and cyclic solar activity. The debris produced by breakups is greatly affected by solar activity, as evidenced by the fragmentation debris curve.

The actual ownership of the satellite population is given in Table 13.1 and its composition illustrated in Fig. 13.2. As can be seen from Fig. 13.3, nearly half of the objects in the catalog have resulted from more than 120 satellite breakups, the history of which is illustrated in Fig. 13.3.[1] The ability to catalog small objects is limited by the power and wavelength of individual radar sites, as well as the limitations on data transmission within the network of radar sites. Consequently, objects smaller than about 10 to 20 cm are not usually cataloged.

The most comprehensive effort to track orbiting objects is carried out by the USSPACECOM. An estimate of USSPACECOM's capability to detect objects in Earth orbit is illustrated in Fig. 13.4 (Refs. 5 and 8). Only the region to the right of the heavy line is accessible to operational radar and optical systems. The

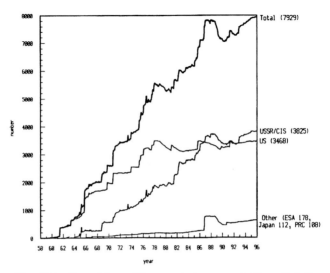

Fig. 13.1   On-orbit satellite population growth (from Ref. 3).

capabilities of the infrared astronomy satellite (IRAS) extend the measurements to smaller objects, as indicated.

The tracked (>10-cm) and estimated (>1-cm) space object densities, as reported in Ref. 7, are illustrated in Fig. 13.5. The peak densities appear at about 800-, 1000-, and 1500-km altitude and are caused by heavy use of these altitudes. A number of high-intensity explosions or breakups of spacecraft and rocket stages have also contributed to the debris population in this environment. The flux of the space objects, or the expected impact per unit area per unit time, is indicated by the right scale in Fig. 13.5.

The inclination distribution of the catalog population is shown in Fig. 13.6. Nearly all of the orbital debris measurements to date show an orbital debris

Table 13.1   Debris and launch watch, 1 April 1995 (Ref. 4)

| Owner | Objects in orbit | | Total |
	Payload	Debris	
USSR/CIS	1320	2488	3808
U.S.	661	2788	3449
Japan	54	55	109
ESA	28	144	172
China	16	92	108
Other	193	33	226
Total	2272	5600	7872

Note: Decayed objects: payload 2381, debris 13291, total 15672. Total number of objects: decayed and in orbit 23544.

APPROXIMATE CATALOG COMPOSITION

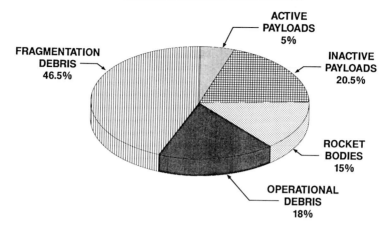

Fig. 13.2   Approximate catalog composition (from Ref. 4).

flux that exceeds the meteoroid flux. These measurements are summarized and compared with the meteoroid flux in Fig. 13.7 (Refs. 5 and 6).

13.4   Space Debris Environment: Geosynchronous Orbit

Because of the uniqueness and usefulness of the geosynchronous equatorial orbit (GEO), the population of objects in this orbit has increased continuously. Reference 7 shows, for example, that at the beginning of 1988, there were 286 cataloged satellites residing in this orbit, not including spent upper stages. Of these, 110–130 operating satellites were on station, along with 150 that were nonfunctional or abandoned.

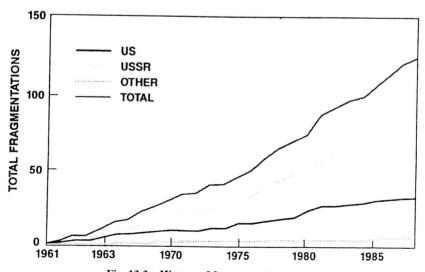

Fig. 13.3   History of fragmentation events.

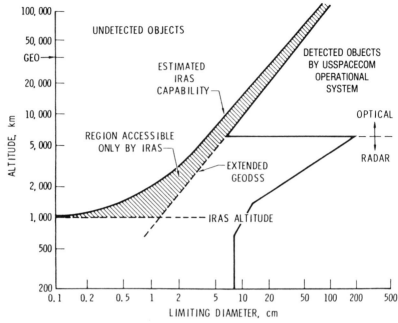

Fig. 13.4 United States capability to detect space objects (from Refs. 5 & 8).

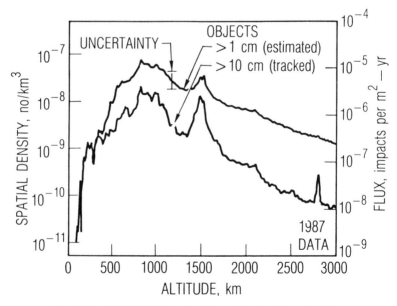

Fig. 13.5 Orbital debris density vs altitude (from Ref. 7).

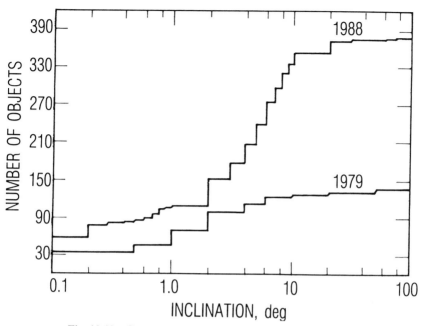

Fig. 13.12   Geosynchronous orbit inclination distribution.

Distributions of the GEO objects with orbital inclination and eccentricity (apogee-perigee difference) are shown in Figs. 13.12 and 13.13, respectively, for the February 1988 and October 1979 catalogs.

A significant increase in the number of objects is apparent in the time interval considered. The drift-rate distribution for the 1988 catalog is also shown in Fig. 13.14. It can be seen that only 70 objects have drift rates of less than 0.01 deg/day and fewer than 180 objects less than 0.1 deg/day. These are "fixed" or active satellites, and the remaining objects are uncontrolled or abandoned debris.

13.5   Spatial Density and Collision Hazard

Spatial Density

The spatial density as a function of radius r and latitude L of the N objects in the geosynchronous ring, defined by a toroidal volume, can be expressed approximately as

$$\rho(r, L) = \sum_{k=1}^{N} \rho_k \qquad (13.1)$$

where

$$\rho_k = \frac{P_k(r_1, r_2) P_k(-L, L)}{\Delta V_k} \qquad (13.2)$$

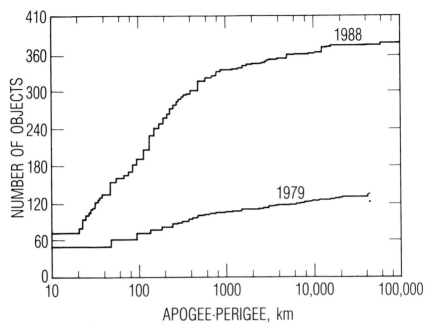

Fig. 13.13   Geosynchronous eccentricity distribution.

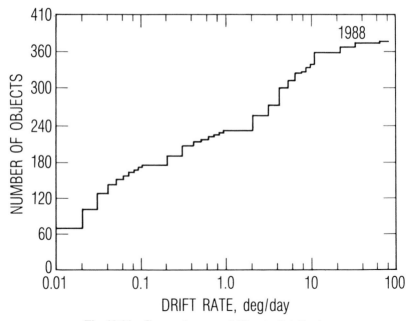

Fig. 13.14   Geosynchronous drift-rate distribution.

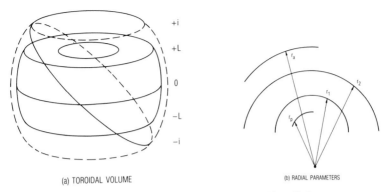

(a) TOROIDAL VOLUME   (b) RADIAL PARAMETERS

Fig. 13.15   Orbital geometry: a) toroidal volume, b) radial parameters.

with the probability of the kth object being within the tortoidal volume,

$$\Delta V_k = \frac{4\pi}{3} \left(r_2^3 - r_1^3 \right) \sin L \qquad (13.3)$$

defined by the arbitrary orbital radii r_1, r_2, and latitude L as shown in Fig. 13.15. Here,

$P_k(r_1, r_2)$ = probability that the kth object is within the radial range Δr
            = $r_2 - r_1$
$P_k(-L, L)$ = probability that the kth object is between the latitudes $-L$ and L

For example, if $r_a \geq r_2 > r_1 \geq r_p$, where r_a and r_p are the orbit apogree and perigee, respectively, and, if $i > L$, where i is the orbit plane inclination of the object,

$$P_k(r_1, r_2) = P_k(r_p, r_2) - P_k(r_p, r_1) \qquad (13.4)$$

where the probability that the object lies between perigee r_p and some radius r, is derived in Ref. 18 as

$$P(r_p, r) = \frac{1}{2} + \frac{1}{\pi} \sin^{-1} \left[\frac{2(r-a)}{r_a - r_p} \right] - \frac{1}{a\pi} \sqrt{(r_a - r)(r - r_p)} \qquad (13.5)$$

and the probability that the object is between the latitudes $-L$ and L is

$$P(-L, L) = \frac{2}{\pi} \sin^{-1} \left(\frac{\sin L}{\sin i} \right) \qquad (13.6)$$

where a and i are the object orbit semimajor axis and inclination, respectively.

Equation (13.1) is plotted in Fig. 13.16 for the 379 objects of the 1988 GEO catalog population considered. The results show that the spatial density is maximum (about 10^{-8} sats/km^3) in a narrow range of the geosynchronous altitude. It decreases by about two orders of magnitude at ±100 km above or below GEO. The subsequent decrease with altitude is not as high as in the first 100-km range, being about three orders of magnitude at ±600-km range.

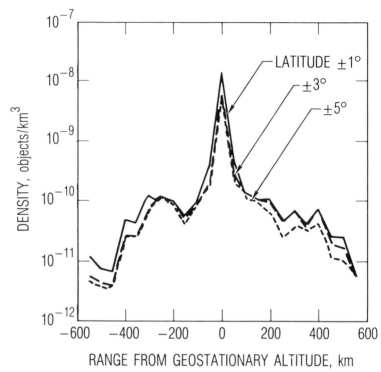

Fig. 13.16   Population density as a function of range and latitude from GEO.

13.6   Collision Hazards Associated with Orbit Operations

The hazards can be broadly defined in two categories; hazards due to explosion and hazards due to collision. Explosions are a major concern because of the probabilities of collision of other satellites with the debris from the explosion. Explosions can be deliberate or accidental. For example, on 13 November 1986, an Ariane third stage experienced an anomalous explosion in a sun-synchronous orbit and left about 460 trackable fragments.

Collisions are also a major concern because an unplanned collision may damage an active spacecraft and because of the collision probabilities of other satellites with the debris. Like explosions, collisions can be planned and unplanned. A planned collision can be part of an orbital test, such as with the Delta 180 mission, for example. There are no confirmed cases yet of unplanned collisions, although there are some breakups that are suspect. The collision can be with another space-craft (Delta 180), with debris (either trackable or untrackable), or with released objects that can collide on subsequent revolutions.

An example of the damage potential is shown in Fig. 13.17 (Ref. 5), which shows the 4-mm-diam crater on the Shuttle window from the STS-7 mission. Energy-discursive x-ray analysis was used to determine the composition of par-tially fused material found in the bottom of the pit. Titanium oxide and small amounts of aluminum, carbon, and potassium were found added to the pit glass.

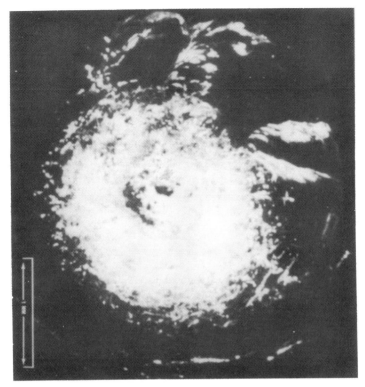

Fig. 13.17   STS-7 window impact.

Crater morphology places the impacting particle diameter at 0.2 mm, with a velocity between 3 and 6 km/s. From these data, it is concluded that the particle was man-made and probably an orbiting paint fleck. This is the first conclusive case in which orbital debris can be shown to have caused operational loss to a space vehicle subsystem.

13.7   Collision Hazard Assessment Methods

Uniform Density

The probability that any two objects will collide is generally a function of the orbital parameters, object size, and time. However, the collision cannot take place unless the orbits approach each other within an effective collision radius. This may occur even for initially nonintersecting orbits because the Earth's oblateness, air drag, and solar-lunar perturbations tend to alter the orbital parameters in time.

A satellite with a projected area A_c, moving with a mean relative velocity $\bar{v}_r$ will sweep out a volume $V = \bar{v}_r A_c \Delta t$ in a time increment Δt. The number of objects encountered is ρV, where ρ is the object density in V. For $\rho V \ll 1$, the probability of collision is approximately

$$p(\text{col}) \approx \rho \bar{v}_r A_c \Delta t \qquad (13.7)$$

Variable Density

For geosynchronous satellites, the collision probability per revolution in an altitude band $\Delta h = h_2 - h_1$ and a latitude band $\Delta\phi = 2\phi$ can be expressed approximately as

$$p(\text{col})/\text{rev} \approx A_c \bar{v}_r T \int_{h_1}^{h_2} \int_{-\phi}^{\phi} \rho(h, \phi) f(h, \phi) \, d\phi \, dh \qquad (13.8)$$

where $\rho(h, \phi)$ is the object density function, $f(h, \phi)$ a weighting function that can be derived from the time spent by a satellite in a latitude band $\Delta\phi$, and T the period of revolution. The mean relative velocity $\bar{v}_r$ is, in general, a function of altitude h and latitude ϕ but may, for simplicity, be regarded as constant for any given Δh and $\Delta\phi$. This approach avoids evaluation of the more precise, but also more complex, path integrals, as has been done in Ref. 10, for example.

As an example, consider the probability that a satellite in GEO will collide with any other satellite in GEO. The probability of collision for a spacecraft in a circular orbit with inclination i, for a given range above or below GEO, has been computed and plotted in Fig. 13.18. The results show that the probability of collision is greatest for GEO satellites and decreases as a function of range above or below GEO. It also increases with inclination because of higher relative velocities between the satellite and the objects in GEO. For example, a decrease of

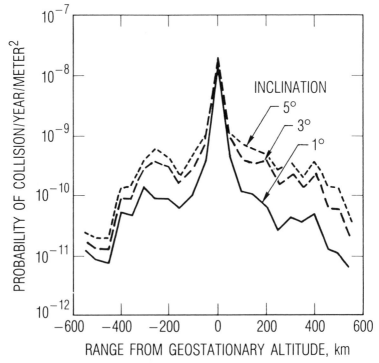

Fig. 13.18 Probability of collision for circular orbits as a function of inclination and range from GEO.

two orders of magnitude is apparent for a spacecraft in an orbit with an inclination of 1 deg at ±100 km above or below GEO.

Distance of Closest Approach

For satellites in mutually inclined circular orbits with equal, or nearly equal, periods of revolution, the distance of closest approach R_{min} occurs in the vicinity of the nodal axis on the ascending or descending passes. R_{min} generally depends on the synodic period or the angular increment $\Delta u = u_1|(T_2 - T_1)|$ where u_1 is the mean motion of satellite 1 and T_1, T_2 are the respective periods of revolution for geosynchronous satellites. Δu is typically a fraction of a degree.

If it is assumed that the position uncertainties associated with the three dimensions (coordinates) of the nominal miss distance R_{min} are Gaussian (normal), with zero biases and equal variance, and are uncorrelated, a bivariate normal density function $f(x, y)$ can be defined in plane xy containing R_{min} that is oriented normal to the relative velocity vector at encounter. Thus,

$$f(x, y) = (2\pi\sigma^2)^{-1} \exp[-x^2/(2\sigma^2)] \exp[-y^2/(2\sigma^2)] \qquad (13.9)$$

A collision can take place only in a region R defined by

$$X_{min} - R_s \leq x \leq X_{min} + R_s$$
$$Y_{min} - R_s \leq y \leq Y_{min} + R_s \qquad (13.10)$$

where X_{min}, Y_{min} define the magnitude of R_{min}, i.e., $R_{min} = (X_{min}^2 + Y_{min}^2)^{1/2}$ and R_s is the effective collision radius. The probability of collision for $R_s \ll R_{min}$ takes the following form:

$$P(col) = \int\int_R f(x, y)\,dx\,dy \qquad (13.11)$$

$$= (2/\pi)(R_s/\sigma)^2 \exp\left[-R_{min}^2/(2\sigma^2)\right]$$

Equation (13.11) is plotted in Fig. 13.19 with R_{min}/σ as a parameter. The probability $p(col)$ takes its maximum value P_{max} at $\sigma = R_{min}/\sqrt{2}$ with R_{min} as a parameter. A plot of P_{max} vs R_{min}/R_s is shown in Fig. 13.20.

For example, the collision probabilities for several close approaches between OPS 6391 and WESTAR-A geosynchronous satellites are shown in Fig. 13.21, plotted as a function of position uncertainty. The values of σ are generally functions of tracking techniques; but, even for the close approaches examined, the collision probabilities were found to be quite small.[20]

13.8 Examples: Collision Hazards in LEO

As has been mentioned previously, the probability of collision is a function of the spacecraft's size, the orbital altitude, and the period of time the spacecraft will remain in orbit. The orbital debris environment in LEO presents a problem for space operations that involve large spacecraft in orbit or satellites in orbit for long

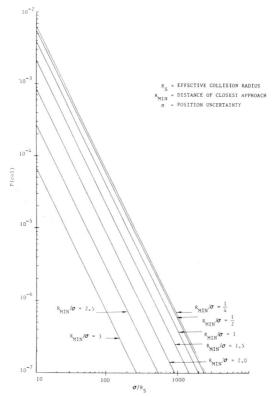

Fig. 13.19   Collision probability as a function of position uncertainty at encounter (from Ref. 11).

periods of time. A space station is the primary example of such a spacecraft and must be shielded over large areas in order to achieve the design safety limits.

The "design driver" is the determination of an acceptable level of risk. For example, the specified level of risk for manned space programs from Apollo to the present has been essentially constant at .005 probability of penetration over the lifetime of the space system. The actual level of risk experienced by these spacecraft has been significantly less than that specified because other design requirements made the spacecraft more robust. The earlier manned space programs addressed only the natural meteoroid environment, but the current Shuttle and proposed Space Station requirements address both the natural meteoroid and the orbital debris environments. Substantial growth of the debris environment may also require additional shielding for smaller satellites.

In order to visualize the implications of orbital debris growth, it is helpful to consider two illustrative cases, as presented in Ref. 2. One is a space station of the general size of the future Space Station, operating at 500 km. The probabilities of impact are approximate, based on equivalent surface area and do not account for directional effects and the relative orientation of component elements. The other is a typical small satellite operating at the LEO most popular satellite altitude of 800 km. For each of these cases, it is instructive to compare the effects of the current

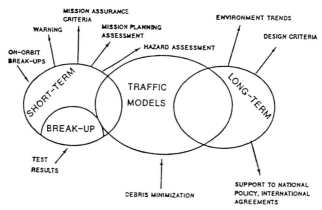

Fig. 13.24   Models.

13.12   Short- and Long-Term Debris Evolution Modeling

The fragmentation of upper stages and satellites accounts for nearly one-half of all trackable debris objects. Understanding the fragmentation process (due to explosion or hypervelocity collision) is therefore essential in order to estimate the mass, number, velocity, and ballistic coefficients of the resultant fragments. The modeling of the fragmentation debris (as orbiting clouds) can be separated into short- and long-term types. *Short-term* may be defined in terms of days after the breakup, while *long-term* implies subsequent evolution of the fragment clouds and the resulting "steady-state" environment due to the breakup. Primary elements of such modeling include: a) determination of breakup causes, b) orbital lifetime of untrackable debris, c) the breakup process (i.e., breakup modeling), d) debris cloud evolution, e) future traffic projections, and f) collision hazards to resident space objects. Figure 13.24 illustrates the interfaces required between various types of models.

13.13   Determination of Breakup Causes

When a breakup event occurs in orbit, it may be due to an explosion caused by residual propellants or by a collision with another object. If explosion-induced, the breakup produces a number of large trackable objects and a smaller population of untrackable debris. This type of breakup probably accounts for most of the fragmentation debris in orbit. Collision-induced satellite fragmentations, on the other hand, are potentially the most threatening to the near-Earth environment. A single hypervelocity collision can produce millions of fragments with diameters of 1 mm or more. Also, the addition of many smaller fragments may not only increase the debris environment but may cause other collisions to occur, resulting in the creation of still more debris. The long-term stability of the debris environment may thus be affected especially in the region from 930- to 1100-km altitude where, for the current environment in the population as a whole, there is a 20% probability per year of collision between two objects of size 1 cm or larger, and a 3.7% probability of catastrophic collision.[27] Reference 28 indicates that the region between 900- and 1000-km altitude is currently unstable, where the debris population will produce fragments from random collisions at a rate that is increasing and greater than the rate of removal by natural forces (e.g., air drag).

13.14 Spacecraft Breakup Modeling

There are a number of empirically based fragmentation models that describe the number, mass, and velocity distributions of fragments resulting from hypervelocity collisions or explosions (see, for example, Refs. 19, 21, 24–26, 29, and 30). There is, however, much uncertainty due to the limited experimental data that must be used to validate such models. Only more testing, coupled with the development of new analytical computational tools, can reduce these uncertainties. The importance of developing a better understanding of the breakup phenomenology for the collisions between LEO objects is also clear. Such collisions are likely to occur with impact velocities in the 8- to 15-km/s range and result in hundreds of trackable fragments and potentially millions of smaller particles. The development of breakup models is essential. These models should include strict energy and momentum conservation laws such as are being implemented in program IMPACT, for example.[34] Other models should be developed by numerical/analytical approaches coupling existing models that describe local area effects due to hypervelocity impact (hydrocodes) with models that describe the spacecraft vibration and deformation due to the impact (structural response codes). Statistical fragmentation models can also be developed using the outputs of other models described previously.

13.15 Debris Cloud Evolution Modeling

The behavior of debris particles following a breakup in orbit must be modeled in order to determine the collision hazard for resident space objects in the vicinity of the breakup. Short-term behavior (measured in days) of a debris cloud for a hypervelocity collision is, for example, described in Refs. 21, 24, and 31. The debris cloud dynamics model in Ref. 21 was developed based on the linearized rendezvous equations for relative motion in orbit. Subsequently, the short-term behavior and modeling of debris clouds in eccentric orbits were developed, which described the shape, in-track particle density variation, and volume of a debris cloud resulting from an isotropic breakup.[31] A typical cloud density profile is shown in Fig. 13.25.

The transition from short-term to long-term (steady-state) must also be modeled. Long-term modeling, which has been an area of research at NASA over a period of years, has contributed to the development of the debris environment models used for the design of the Space Station and other spacecraft.[32]

13.16 Lifetime of Nontrackable Debris

The long-term debris environment in Earth orbit may be expressed in terms of the source and sink effects controlling the environment. Sources are all new objects, including fragments resulting from explosions or collisions of objects in orbit. Sink effects govern the removal of objects such as the atmospheric drag that tends to keep the debris population at low altitude nearly constant.

First evidence of the existence of small debris in orbit resulted from examining the Skylab IV/Apollo windows,[33] which showed aluminum-lined pits in about half

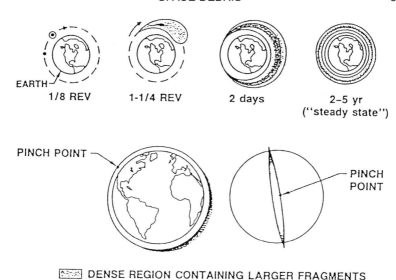

EARTH
1/8 REV   1-1/4 REV   2 days   2-5 yr
                                            ("steady state")

PINCH POINT   PINCH
                                               POINT

▨ DENSE REGION CONTAINING LARGER FRAGMENTS

Fig. 13.25   Evolution of a debris cloud.

of the hypervelocity pits found there. Other experiments, such as the Explorer 46 Meteor Bumper Experiment, showed evidence of debris smaller than 0.1 mm.[22] Much good data have also been obtained from the returned Solar-Max insulation blankets and aluminum louvers. These data showed that the man-made debris dominated the micrometeoroid particle flux for sizes smaller than about 0.01 mm. After nearly six years in space, the Long-Duration Exposure Facility (LDEF) results have also shown that there were a total of more than 34,000 impacts on the satellite following its recovery from orbit in January 1990 by the orbiter Columbia. Of these, more than 3000 impacts were in the 0.5- to 5-mm size range. Several studies of the spacecraft surfaces returned from LDEF are reported in Refs. 36 and 37. Orbital decay time vs initial altitude is shown in Fig. 13.26.

13.17   Methods of Debris Control

In view of the increasing threat of space debris to operations in orbit, NASA, the Department of Defense, and several space agencies from Russia, Japan, and Europe are participating jointly in the attempt to minimize the threat in the future. Reference 35, for example, outlined a number of debris-control measures which, if implemented, could decrease the quantity of space debris by prevention or debris removal. These control options fall into three categories: those requiring minimal impact on operations, those requiring changes in hardware or operations, and those requiring technology development. Options in the first category recommended for immediate application are:

1) No deliberate breakups of spacecraft, which produce debris in long-lived orbits.

2) Minimization of mission-related debris.

3) Safing procedures for all rocket bodies and spacecraft that remain in orbit after completion of their mission.

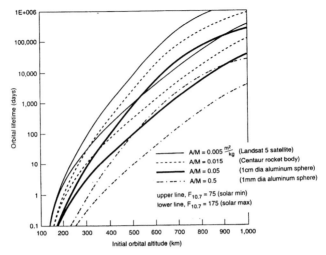

Fig. 13.26   Orbital decay time vs altitude (from Ref. 41).

4) Selection of transfer orbit parameters to ensure the rapid decay of transfer stages.

5) Reorbiting of geosynchronous equatorial satellites at the end of life (minimum altitude increase, 300–400 km).

6) Upper-stage and separated apogee kick motors used for geostationary satellites should be inserted into a disposal orbit at least 300 km above the geostationary orbit.

Second category options aim for removing upper stages and dead spacecraft from orbit. This can be accomplished with deorbiting maneuvers to ensure atmospheric entry over ocean areas. Debris-control options in category three require new developments in which, in general, technical feasibility and cost-effectiveness must be demonstrated. Installation of drag-enhancement devices, use of lasers, debris sweeps, and so forth, fall into this category.

Once a better understanding of the debris environment is gained, national and international agreements should be reached to control the space debris problem and ensure safe future use of space for all mankind. For additional information on the subject of space debris, see Refs. 36–47.

References

[1]Johnson, N. L., and McKnight, D. S., *Artificial Space Debris,* Krieger Publishing, Malabar, FL, 1987.

[2] "Report on Orbital Debris," Interagency Group (Space) for National Security Council, Washington, DC, Feb. 1989.

[3] "Interagency Report on Orbital Debris," Office of Science and Technology Policy, 1995.

[4]McKnight, D. S. (ed.), *Orbital Debris Monitor,* Vol. 8, Apr. 1995.

[5]Kessler, D. J., "Orbital Debris-Technical Issues," presentation to the USAF Scientific Advisory Board on Space Debris, Jan. 1987.

[6]Atkinson, D. R., Watts, A. J., and Crowell, L., "Spacecraft Microparticle Impact Flux Definition," Final Report for University of California, Lawrence Livemore National Laboratory, UCRL-RC-108788, Aug. 30, 1991.

[7] "Space Debris," Space Debris Working Group, European Space Agency, ESTEC, Noordwijk, Netherlands, Report. SP 1109, Nov. 1988.

[8]Chobotov, V. A., and Wolfe, M. G., "The Dynamics of Orbiting Debris and the Impact on Expanded Operations in Space," *Journal of the Astronautical Sciences,* Vol. 38, Jan.–March 1990.

[9]Perek, L., "The Scientific and Technical Aspects of the Geostationary Orbit," 38th IAF Congress, IAA Paper 87-635, 1987.

[10]Hechler, M., and Vanderha, J. C., "Probability of Collisions in the Geostationary Ring," *Journal of Spacecraft and Rockets,* Vol. 18, July–Aug. 1981.

[11]Chobotov, V. A., "The Collision Hazard in Space," *Journal of the Astronautical Sciences,* July–Sept. 1982, pp. 191–212.

[12]Chobotov, V. A., "Classification of Orbits with Regard to Collision Hazard in Space," *Journal of Spacecraft and Rockets,* Vol. 20, Sept.–Oct. 1983, pp. 484–490.

[13]Perek, L., "Safety in the Geostationary Orbit After 1988," 40th IAF Congress, IAF Paper 89-632, 1989.

[14]Bird, A. G., "Special Considerations for GEO-ESA," AIAA Paper 90-1361, Baltimore, MD, 1990.

[15]Fenoglio, L., and Flury, W., "Long-Term Evolution of Geostationary and Near-Geostationary Orbits," ESA/ESOC, Darmstadt, Germany, MAS Working Paper 260-1987.

[16]Chobotov, V. A., "Disposal of Spacecraft at End of Life in Geosynchronous Orbit," *Journal of Spacecraft and Rockets,* Vol. 27, No. 4, 1990, pp. 433–437.

[17]Yasaka, T., and Oda, S., "Classification of Debris Orbits With Regard to Collision Hazard in Geostationary Region," 41st IAF Congress, AIAA Paper 90-571, 1990.

[18]Dennis, N. G., "Probabilistic Theory and Statistical Distribution of Earth Satellites," *Journal of British Interplanetary Society,* Vol. 25, 1972, pp. 333–376.

[19]Su, S. Y., and Kessler, D. J., "Contribution of Explosion and Future Collision Fragments to the Orbital Debris Environment," COSPAR, Graz, Austria, June 1984.

[20]Chobotov, V. A., and Johnson, C. G., "Effects of Satellite Bunching on the Probability of Collision in Geosynchronous Orbit," *Journal of Spacecraft and Rockets* (to be published).

[21]Chobotov. V. A., "Dynamics of Orbital Debris Clouds and the Resulting Collision Hazard to Spacecraft," *Journal of the British Interplanetary Society,* Vol. 43, May 1990, pp. 187–195.

[22]Kessler, D. J., Reynolds, R. C., and Anz-Meador, P. D., "Orbital Debris Environment for Spacecraft Designed to Operate in Low Earth Orbit," NASA TM 100-471, April 1988.

[23]Eichler, P., and Rex, D., "Debris Chain Reactions," AIAA Paper 90-1365, April 1990.

[24]Chobotov, V. A., et al., "Dynamics of Debris Motion and the Collision Hazard to Spacecraft Resulting from an Orbital Breakup," The Aerospace Corp., El Segundo, CA.

[25]Chobotov, V. A., and Spencer, D. B., "Debris Evolution and Lifetime Following an Orbital Breakup," *Journal of Spacecraft and Rockets,* Vol. 28, No. 6, Nov.–Dec., 1991, pp. 670–676.26.

[26]Reynolds, D. C., "A Review of Orbital Debris Environment Modeling at NASA/JSC," AIAA/NASA/DOD Orbital Debris Conference, Baltimore, MD, AIAA Paper 90-1355, April 1990.

[27]Eichler, P., and Rex, D., "Debris Chain Reactions," AIAA/NASA/DOD Orbital Debris Convergence, AIAA Paper 90-1365, Baltimore, MD, April 1990.

[28]Kessler, D. J., "Collisional Cascading: The Limits of Population Growth in Low Earth Orbit," *Advances in Space Research* (COSPAR, 1990).

[29]Reynolds, R. C., and Potter, A. E., Jr., "Orbital Debris Research at NASA Johnson Space Center, 1986–1988," NASA Technical Memorandum 102-155, Sept.1989.

[30]McKnight, D. S., "Determination of Breakup Initial Conditions," 29th Aerospace Sciences Meeting, AIAA Paper 91-0299, Reno, NV, Jan. 1991.

[31]Jenkin, A. B., "DEBRIS: A Computer Program for Debris Cloud Modeling," 44th Congress of IAF, Oct. 16–22, 1993, Graz, Austria (AIAA 6.3-93-746).

[32]Kessler, D. J., and Reynolds, R. C., "Orbital Debris Environment for Spacecraft Designed to Operate in Low Earth Orbit," NASA TM 100471, April 1989.

[33]Clanton, U. S., Zook, H. A., and Schultz, R. A., "Hypervelocity Impacts on Skylab IV/Apollo Windows," NASA CP 2360, pp. 177–189, 1985.

[34]Sorge M. E., and Johnson, C. G., Space Debris Hazard Software: Program Impact Version 3.0 User's Guide, Aerospace Corp. TOR-93(3076)-3, Aug. 1993.

[35] "Position Paper on Orbital Debris," *International Academy of Astronautics,* March 8, 1993.

[36] *Proceedings of the First European Conference on Space Debris,* Darmstadt, Germany, April 5–7, 1993, ESA SD-01.

[37]McDonnell, J.A.M. (ed.), *Hypervelocity Impacts in Space,* University of Kent, England, 1992.

[38]Portree, D.S.F., and Loftus, J. P., Jr., "Orbital Debris and Near-Earth Environmental Management: A chronology," NASA.

[39]Flury, W., (ed.), "Space Debris," *Advances in Space Research,* Vol. 13, No. 8, 1992.

[40]Toda, S., "Recent Space Debris Activities in Japan," *Earth Space Review,* Vol. 4, No. 3, 1995.

[41] *Orbital Debris—A Technical Assessment,* National Academy Press, 1995.

[42]Smirnov, N. N., Lebedev, V. V., and Kiselev, A. B., "Mathematical Modeling of Space Debris Evolution in Low Earth Orbit," 19th ISTS, Yokohama, Japan, May 1994.

[43]Nazarenko, A. I., "Prediction and Analysis of Orbital Debris Environment Evolution," First European Conference on Space Debris, Darmstadt, Germany, Apr. 1993.

[44]Chernyavskiy, A. G., Chernyavskiy, G. M., Johnson, N., and McKnight, D., "A Simple Case of Space Environmental Effects," 44th International Astronautical Federation Congress, Graz, Austria, Oct. 1993.

[45]Maclay, T. D., Madler, R. A., McNamara, R., and Culp, R. D., "Orbital Debris Hazard Analysis for Long Term Space Assets," *Proceedings of the Workshop on Hypervelocity Impacts in Space,* Univ. of Kent, Canterbury, U. K., July 1991.

[46]Culp, R. D. et al,, "Orbital Debris Studies at the University of Colorado," First European Conference on Space Debris, Darmstadt, Germany, Apr. 1993.

[47]Veniaminov, S. S., "The Methods and Experience of Detecting Small and Weakly Contrasting Space Objects," First European Conference on Space Debris, Darmstadt, Germany, Apr. 1993.

14
Optimal Low-Thrust Orbit Transfer

14.1 Introduction

The theory of optimal low-thrust orbit transfer has received a great deal of attention in the astrodynamics and flight mechanics literature over the past several decades. This chapter begins with a detailed description of some fundamental analytic results obtained by Edelbaum, which are widely in use by the practitioners in the aerospace industry. The reader, after becoming familiar with the simplified transfer analysis, is invited to consider the treatment of the exact transfer problem in the subsequent sections. Drawing on the pioneering work of the Americans Broucke, Cefola, and Edelbaum, who perfected the theory of orbital mechanics in terms of nonsingular orbital elements, examples of optimal orbit transfers are generated and discussed, and all the relevant equations needed to develop unconstrained orbit transfer computer codes are exposed and derived.

14.2 The Edelbaum Low-Thrust Orbit-Transfer Problem

A discussion of the problem of optimal low-thrust transfer between inclined circular orbits was presented by Edelbaum in the early 1960s.[1] Assuming constant acceleration and constant thrust vector yaw angle within each revolution, Edelbaum linearizes the Lagrange planetary equations of orbital motion about a circular orbit and, using the velocity as the independent variable, reduces the transfer optimization problem to a problem in the theory of maxima. The variational integral involves a single constant Lagrange multiplier since it involves a single integral constraint equation for the transfer time or velocity change while maximizing the change in orbital inclination. The control variable being the yaw angle, the necessary condition for a stationary solution is obtained by simply setting the partial derivative of the integrand of the variational integral with respect to the control to zero. This optimum control is then used in the right-hand sides of the original equations of motion, which are integrated analytically to provide expressions for the time and inclination in terms of the independent variable, the orbital velocity. Two expressions for the inclination are provided to cover the case of large inclination change transfers. This complication arises if orbital velocity is adopted as the independent variable. However, a single expression for the inclination change can be obtained that is uniformly valid throughout any desired transfer if the original Edelbaum problem is cast into a minimum-time transfer problem using the more direct formalism of optimal control theory. Following is a discussion of Edelbaum's original analysis, as well as the formulation using optimal control theory.

Edelbaum's Analysis

The full set of the Gaussian form of the Lagrange planetary equations for near-circular orbits is given by

$$\dot{a} = \frac{2af_t}{V} \tag{14.1}$$

$$\dot{e}_x = \frac{2 f_t c_\alpha}{V} - \frac{f_n s_\alpha}{V} \tag{14.2}$$

$$\dot{e}_y = \frac{2 f_t s_\alpha}{V} + \frac{f_n c_\alpha}{V} \tag{14.3}$$

$$\dot{i} = \frac{f_h c_\alpha}{V} \tag{14.4}$$

$$\dot{\Omega} = \frac{f_h s_\alpha}{V s_i} \tag{14.5}$$

$$\dot{\alpha} = n + \frac{2 f_n}{V} - \frac{f_h s_\alpha}{V \tan i} \tag{14.6}$$

where s_α and c_α stand for $\sin\alpha$ and $\cos\alpha$, respectively, and a stands for the orbit semimajor axis, i for inclination, and Ω for the right ascension of the ascending node; $e_x = e\cos\omega$, and $e_y = e\sin\omega$, with e and ω standing for orbital eccentricity and argument of perigee. Finally, $\alpha = \omega + M$ represents the mean angular position, M the mean anomaly, and $n = (\mu/a^3)^{1/2}$ the orbit mean motion, with μ standing for the Earth gravity constant. For near-circular orbits, $V = na = (\mu/a)^{1/2}$. The components of the thrust acceleration vector along the tangent, normal, and out-of-plane directions are depicted by f_t, f_n, and f_h, with the normal direction oriented toward the center of attraction. If we assume only tangential and out-of-plane acceleration, and that the orbit remains circular during the transfer, the Eqs. (14.1–14.6) reduce to

$$\dot{a} = \frac{2 a f_t}{V} \tag{14.7}$$

$$\dot{i} = \frac{f_h c_\alpha}{V} \tag{14.8}$$

$$\dot{\Omega} = \frac{f_h s_\alpha}{V s_i} \tag{14.9}$$

$$\dot{\alpha} = n - \frac{f_h s_\alpha}{V \tan i} \tag{14.10}$$

If f represents the magnitude of the acceleration vector, and β the out-of-plane or thrust yaw angle, then $f_t = f c_\beta$ and $f_h = f s_\beta$. Furthermore, $\alpha = \omega + M = \omega + \theta^* = \theta$, the angular position when $e = 0$, with $\theta = nt$ and θ^* the true anomaly. If the angle β is held piecewise constant switching sign at the orbital antinodes, then the $f_h s_\alpha$ terms above in Eqs. (14.9) and (14.10) will have a net zero contribution such that the system of differential equations further reduces to

$$\dot{a} = \frac{2 a f_t}{V} \tag{14.11}$$

$$\dot{i} = \frac{c_\theta f_h}{V} \tag{14.12}$$

$$\dot{\theta} = n \tag{14.13}$$

We can now average out the angular position θ in Eq. (14.12) by integrating with respect to θ and by holding f, β, and V constant

$$\int_0^{2\pi} \left(\frac{di}{dt}\right) d\theta = \frac{2fs_\beta}{V} \int_{-\pi/2}^{\pi/2} c_\theta d\theta$$

$$2\pi \frac{di}{dt} = \frac{4fs_\beta}{V}$$

$$\frac{di}{dt} = \frac{2fs_\beta}{\pi V} \tag{14.14}$$

From the energy equation $V^2/2 - \mu/r = -\mu/2a$, with $r = a$, and with Eq. (14.11) used to eliminate the semimajor axis,

$$dV = -\left[\frac{\mu}{2Va^2}\right] da$$

$$= -fc_\beta dt$$

$$\frac{dV}{dt} = -fc_\beta \tag{14.15}$$

Equation (14.14) can also be obtained by dividing Eq. (14.12) by Eq. (14.13),

$$\frac{di}{d\theta} = \frac{c_\theta f_h}{V n}$$

$$\Delta i = \frac{2f_h}{V n} \int_{-\pi/2}^{\pi/2} c_\theta d\theta = \frac{4f_h}{V n}$$

and, since $\Delta t = 2\pi a/V$,

$$\frac{di}{dt} = \frac{\Delta i}{\Delta t} = \frac{2f_h}{V\pi}$$

Equations (14.14) and (14.15) can be replaced by the following set, where V is now the independent variable,

$$\frac{di}{dV} = -\frac{2\tan\beta}{\pi V} \tag{14.16}$$

$$\frac{dt}{dV} = -\frac{1}{fc_\beta} \tag{14.17}$$

Let I represent the functional to be maximized,

$$I = \int_{V_0}^{V_f} \left(\frac{di}{dV}\right) dV = -\int_{V_0}^{V_f} \frac{2}{\pi V} \tan\beta \, dV \tag{14.18}$$

and let J represent the integral constraint given by

$$J = \int_{V_0}^{V_f} \left(\frac{dt}{dV} \right) dV = \text{const} \tag{14.19}$$

Let us adjoin J to I by way of a constant Lagrange multiplier λ such that the optimization problem is now reduced to a succession of ordinary maximum problems for each value of V between V_0 and V_f, the initial and final velocities, respectively. The necessary condition for a stationary solution of the augmented integral,

$$K = I + \lambda J = \int_{V_0}^{V_f} \left[-\frac{2}{\pi V} \tan \beta - \frac{\lambda}{f c_\beta} \right] dV \tag{14.20}$$

is then simply given by

$$\frac{\partial}{\partial \beta} \left[\frac{2}{\pi V} \tan \beta + \frac{\lambda}{f c_\beta} \right] = 0 \tag{14.21}$$

The optimization problem consists, therefore, of the maximization of the inclination change subject to the constraint of given total transfer time since

$$\Delta V = f t \tag{14.22}$$

This constraint is equivalent to the fixed ΔV constraint for constant acceleration f. Furthermore, V_0 and V_f being given, the initial and final radii are, therefore, given too since the orbits are assumed circular. With the acceleration being applied continuously, this problem is equivalent to minimizing the total transfer time for a given change in the inclination and velocity. This is also equivalent to minimizing the total ΔV or propellant usage because the thrust is always on and no coasting arcs are allowed. In this latter case, I and J are simply interchanged to yield the optimality condition

$$\frac{\partial}{\partial \beta} \left[\frac{1}{f c_\beta} + \lambda_i \frac{2}{\pi V} \tan \beta \right] = 0 \tag{14.23}$$

From Eq. (14.21), it follows that

$$V s_\beta = -\frac{2f}{\pi \lambda} = \text{const} = V_0 s_{\beta_0} \tag{14.24}$$

$$\lambda = -\frac{2f}{\pi V_0 s_{\beta_0}} \tag{14.25}$$

The optimal β steering law given by Eq. (14.24) can be used in Eq. (14.17) for dV/dt in order to obtain the expression for the velocity as a function of time t or $\Delta V = f t$.

$$\frac{dV}{dt} = -f c_\beta$$

$$f \, dt = -\frac{dV}{c_\beta} = \frac{-dV}{\pm\left(1 - s_\beta^2\right)^{1/2}}$$

$$\int_0^t f \, dt = -\int_{V_0}^V \frac{V \, dV}{\pm\left(V^2 - V_0^2 s_{\beta_0}^2\right)^{1/2}}$$

$$\Delta V = ft = -\frac{1}{\pm}\left[\left(V^2 - V_0^2 s_{\beta_0}^2\right)^{1/2} - (\pm)V_0 c_{\beta_0}\right]$$

$$\Delta V = V_0 c_{\beta_0} \mp \left(V^2 - V^2 s_\beta^2\right)^{1/2} = V_0 c_{\beta_0} \mp (\pm)V c_\beta \qquad (14.26)$$

$$\Delta V = V_0 c_{\beta_0} - V c_\beta \qquad (14.27)$$

From Eq. (14.26),

$$\Delta V - V_0 c_{\beta_0} = \mp \left(V^2 - V^2 s_\beta^2\right)^{1/2} = \mp \left(V^2 - V_0^2 s_{\beta_0}^2\right)^{1/2}$$

and, after squaring,

$$V^2 = V_0^2 + \Delta V^2 - 2\Delta V V_0 c_{\beta_0} \qquad (14.28)$$

This then represents V as a function of time since $\Delta V = ft$. The initial yaw angle β_0 must still be determined. In a similar way, Eq. (14.16) for di/dV can be integrated to provide an expression for the evolution of the inclination in time.

$$\frac{di}{dV} = -\frac{2}{\pi V}\tan\beta = -\frac{2}{\pi V}\frac{s_\beta}{c_\beta} = -\frac{2}{\pi V}\frac{V s_\beta}{\left(V^2 - V^2 s_\beta^2\right)^{1/2}}$$

such that, with the use of $V s_\beta = V_0 s_{\beta_0}$,

$$\int_{i_0}^i di = -\frac{2}{\pi}V_0 s_{\beta_0}\int_{V_0}^V \frac{dV}{V\left(V^2 - V_0^2 s_{\beta_0}^2\right)^{1/2}}$$

$$\Delta i = -\frac{2}{\pi}\sin^{-1}\left(\frac{-V_0 s_{\beta_0}}{V}\right)\Bigg|_{V_0}^V$$

$$\Delta i = -\frac{2}{\pi}\left[\sin^{-1}(s_{\beta_0}) - \sin^{-1}\left(\frac{V_0 s_{\beta_0}}{V}\right)\right]$$

$$\Delta i = -\frac{2}{\pi}\beta_0 + \frac{2}{\pi}\sin^{-1}\left(\frac{V_0 s_{\beta_0}}{V}\right) \qquad (14.29)$$

and, since $V_0 s_{\beta_0} = V s_\beta$,

$$\Delta i = -\frac{2}{\pi}(\beta - \beta_0)$$

Now, since the inverse sine function in Eq. (14.29) is double-valued in the interval $(0, 2\pi)$, it is necessary to write this function as

$$\sin^{-1}\left(\frac{V_0 s_{\beta_0}}{V}\right) \text{ if } \sin^{-1}\left(\frac{V_0 s_{\beta_0}}{V}\right) < \frac{\pi}{2}$$

and

$$\frac{\pi}{2} + \left[\frac{\pi}{2} - \sin^{-1}\left(\frac{V_0 s_{\beta_0}}{V}\right)\right] = \pi - \sin^{-1}\left(\frac{V_0 s_{\beta_0}}{V}\right) \quad \text{if} \quad \sin^{-1}\left(\frac{V_0 s_{\beta_0}}{V}\right) > \frac{\pi}{2}$$

since the function is symmetrical with respect to $\pi/2$. In the second of the preceding conditions, Δi can be written as

$$\Delta i = \frac{2}{\pi}\left[\pi - \sin^{-1}\left(\frac{V_0 s_{\beta_0}}{V}\right)\right] - \frac{2}{\pi}\beta_0 \tag{14.30}$$

or

$$\Delta i = 2 - \frac{2}{\pi}\sin^{-1}\left(\frac{V_0 s_{\beta_0}}{V}\right) - \frac{2}{\pi}\beta_0 \tag{14.31}$$

This is equivalent to writing Eqs. (14.29) and (14.31) as

$$\Delta i = \frac{2}{\pi}(\beta - \beta_0) \quad \text{if } \beta < \frac{\pi}{2} \tag{14.32}$$

$$\Delta i = 2 - \frac{2}{\pi}(\beta + \beta_0) \quad \text{if } \beta > \frac{\pi}{2} \tag{14.33}$$

Of course, the 2 in Eq. (14.33) is given in radians, and it corresponds to 114.6 deg. Finally, from Eq. (14.26),

$$\Delta V = V_0 c_{\beta_0} - \left(V^2 - V_0^2 s_{\beta_0}^2\right)^{1/2} \quad \text{if } \Delta V - V_0 c_{\beta_0} < 0 \tag{14.34}$$

$$\Delta V = V_0 c_{\beta_0} + \left(V^2 - V_0^2 s_{\beta_0}^2\right)^{1/2} \quad \text{if } \Delta V - V_0 c_{\beta_0} > 0 \tag{14.35}$$

From $\Delta V = V_0 c_{\beta_0} - V c_\beta$, the condition $\Delta V - V_0 c_{\beta_0} < 0$ is identical to $c_\beta > 0$ or $\beta < \pi/2$ or $\sin^{-1}(V_0 s_{\beta_0}/V) < \pi/2$, and the condition $\Delta V - V_0 c_{\beta_0} > 0$ is identical to $\beta > \pi/2$ or $\sin^{-1}(V_0 s_{\beta_0}/V) > \pi/2$ such that the Edelbaum analysis leads to the following set of equations:

1) If $\Delta V - V_0 c_{\beta_0} < 0$, then,

$$\left.\begin{aligned} V &= \left(V_0^2 - 2V_0\Delta V c_{\beta_0} + \Delta V^2\right)^{1/2} \\ \Delta i &= \frac{2}{\pi}\sin^{-1}\left(\frac{V_0 s_{\beta_0}}{V}\right) - \frac{2}{\pi}\beta_0 = \frac{2}{\pi}(\beta - \beta_0) \end{aligned}\right\} \tag{14.36}$$

2) If $\Delta V - V_0 c_{\beta_0} > 0$, then,

$$\left.\begin{aligned} V &= \left(V_0^2 - 2V_0\Delta V c_{\beta_0} + \Delta V^2\right)^{1/2} \\ \Delta i &= 2 - \frac{2}{\pi}\sin^{-1}\left(\frac{V_0 s_{\beta_0}}{V}\right) - \frac{2}{\pi}\beta_0 = 2 - \frac{2}{\pi}(\beta + \beta_0) \end{aligned}\right\} \tag{14.37}$$

The preceding equations show that one must monitor the condition $\Delta V - V_0 c_{\beta_0}$ and use Eq. (14.36) to describe the transfer starting from time 0 and later switch to Eqs. (14.37) as soon as $t = \Delta V / f$ exceeds $V_0 c_{\beta_0}/f$, which will take place for large transfers as will be shown later by an example. For large transfers, ΔV as given in Eq. (14.26) could become double-valued in V such that one must use $\Delta V = V_0 c_{\beta_0} - (V^2 - V_0^2 s_{\beta_0}^2)^{1/2}$ from V_0 to $V_{0} s_{\beta_0}$ and $\Delta V = V_0 c_{\beta_0} + (V^2 - V_0^2 s_{\beta_0}^2)^{1/2}$ from $V = V_{0} s_{\beta_0}$ to V_f, where $V_{0} s_{\beta_0} < V_f$. This minimum velocity takes place when $\Delta V = V_0 c_{\beta_0}$, indicating that the orbit will grow to become larger than the final desired orbit and later shrink to that desired orbit. This will happen when larger inclination changes are required since then the orbit plane rotation will be carried out mostly at those higher intermediate altitudes. This, of course, is the result of the trade between inclination and radius or velocity. From

$$\Delta V = V_0 c_{\beta_0} \mp \left(V^2 - V_0^2 s_{\beta_0}^2\right)^{1/2}$$

we have

$$\frac{\partial \Delta V}{\partial V} = \mp \frac{V}{\left(V^2 - V_0^2 s_{\beta_0}^2\right)^{1/2}}$$

which is equal to ∞ for $V = V_{0} s_{\beta_0}$, the minimum velocity reached. The initial yaw angle β_0 can be obtained from the terminal conditions at time t_f. At $t = t_f$, $V = V_f$ and $\Delta i = \Delta i_f$. Using Eq. (14.36) for Δi, we get

$$\frac{\pi}{2} \Delta i_f + \beta_0 = \sin^{-1}\left(\frac{V_{0} s_{\beta_0}}{V_f}\right)$$

$$\sin\left(\beta_0 + \frac{\pi}{2}\Delta i_f\right) = \frac{V_{0} s_{\beta_0}}{V_f}$$

$$\sin \beta_0 \cos \frac{\pi}{2}\Delta i_f + \cos \beta_0 \sin \frac{\pi}{2}\Delta i_f = \frac{V_{0} s_{\beta_0}}{V_f}$$

Dividing both sides by c_{β_0} yields

$$\left[\cos \frac{\pi}{2}\Delta i_f - \frac{V_0}{V_f}\right] \tan \beta_0 = -\sin \frac{\pi}{2}\Delta i_f$$

$$\tan \beta_0 = \frac{\sin \frac{\pi}{2}\Delta i_f}{\frac{V_0}{V_f} - \cos \frac{\pi}{2}\Delta i_f} \tag{14.38}$$

Now, carrying out the same manipulations using Eq. (14.37), we get

$$\frac{\pi}{2}\left(\Delta i_f - 2 + \frac{2}{\pi}\beta_0\right) = -\sin^{-1}\left(\frac{V_{0} s_{\beta_0}}{V_f}\right)$$

$$\pi - \left(\beta_0 + \frac{\pi}{2}\Delta i_f\right) = \sin^{-1}\left(\frac{V_{0} s_{\beta_0}}{V_f}\right)$$

$$\sin\left(\beta_0 + \frac{\pi}{2}\Delta i_f\right) = \frac{V_0 s_{\beta_0}}{V_f}$$

$$\sin\beta_0 \cos\frac{\pi}{2}\Delta i_f + c_{\beta_0}\sin\frac{\pi}{2}\Delta i_f = \frac{V_0 s_{\beta_0}}{V_f}$$

Dividing by c_{β_0}

$$\tan\beta_0 = \frac{\sin\frac{\pi}{2}\Delta i_f}{\frac{V_0}{V_f} - \cos\frac{\pi}{2}\Delta i_f} \qquad (14.39)$$

Equations (14.38) and (14.39) indicate that β_0 is given by

$$\beta_0 = \tan^{-1}\left[\frac{\sin\frac{\pi}{2}\Delta i_f}{\frac{V_0}{V_f} - \cos\frac{\pi}{2}\Delta i_f}\right] \qquad (14.40)$$

regardless of whether $\Delta V - V_0 c_{\beta_0} < 0$ or $\Delta V - V_0 c_{\beta_0} > 0$ and, from $\Delta V = V_0 c_{\beta_0} - V c_\beta$, the yaw angle β at future times is given by

$$\beta = \cos^{-1}\left[\frac{V_0 c_{\beta_0} - \Delta V}{\left(V_0^2 - 2V_0\Delta V c_{\beta_0} + \Delta V^2\right)^{1/2}}\right] \qquad (14.41)$$

where $\Delta V = ft$ and where $0 \le \beta \le \pi$. This expression is better than $V s_\beta = V_0 s_{\beta_0}$, which would yield $\beta = \sin^{-1}(V_0 s_{\beta_0}/V)$ since β could, for large transfers, exceed $\pi/2$.

If the evolution of Δi as a function of time or velocity or ΔV is desired, Eq. (14.36) for Δi is used until $\Delta V = V_0 c_{\beta_0}$. When $\Delta V - V_0 c_{\beta_0} > 0$, Δi as given in Eq. (14.37) is used. However, in Eq. (14.37), the inverse sine function will always return a β angle that is always less than $\pi/2$, and this value for β is the correct value to be used in $\Delta i = 2 - (2/\pi)(\beta + \beta_0)$. This β angle is clearly not the real yaw angle since, in this case, it would be given by $\pi - \beta$ with $\beta < \pi/2$, such that the yaw angle is now larger than $\pi/2$. If the real β angle is used in $\Delta i = 2 - (2/\pi)(\beta + \beta_0)$, we get

$$\Delta i = 2 - \frac{2}{\pi}(\pi - \beta + \beta_0) = \frac{2}{\pi}(\beta - \beta_0) \qquad (14.42)$$

Equation (14.42) is universally valid for all yaw angles $0 \le \beta < 180$ deg and should be the only one used. Equation (14.42) will effectively replace Eqs. (14.36) and (14.37), provided that the angle β is computed from Eq. (14.41). Since the sign of $\Delta V - V_0 c_{\beta_0}$ is effectively accounted for in Eq. (14.41), it will return the yaw angle β to be used in Eq. (14.42) for the unambiguous evaluation of Δi. Expressions for β_0 and β can also be obtained by using the identity in Eq. (14.42) since

$$\cos\frac{\pi}{2}\Delta i = c_\beta c_{\beta_0} + s_\beta s_{\beta_0} \qquad (14.43)$$

$$\sin \frac{\pi}{2} \Delta i = s_\beta c_{\beta_0} - s_{\beta_0} c_\beta \qquad (14.44)$$

If we multiply Eq. (14.43) by VV_0 and replace Vs_β by $V_0 s_{\beta_0}$ and Vc_β by $V_0 c_{\beta_0} - \Delta V$, we get, after regrouping terms,

$$c_{\beta_0} = \frac{V_0 - V \cos \frac{\pi}{2} \Delta i}{\Delta V} \qquad (14.45)$$

In a similar manner, from Eq. (14.43), if we replace this time $V_0 s_{\beta_0}$ by Vs_β and $V_0 c_{\beta_0}$ by $\Delta V + Vc_\beta$, we get an expression for c_β:

$$c_\beta = \frac{V_0 \cos \frac{\pi}{2} \Delta i - V}{\Delta V} \qquad (14.46)$$

Equation (14.44) can also be written as

$$V V_0 \sin \frac{\pi}{2} \Delta i = V s_\beta V_0 c_{\beta_0} - V_0 s_{\beta_0} V c_\beta$$

If we replace Vs_β by $V_0 s_{\beta_0}$ and $V_0 c_{\beta_0}$ by $\Delta V + Vc_\beta$, then

$$s_{\beta_0} = \frac{V \sin \frac{\pi}{2} \Delta i}{\Delta V} \qquad (14.47)$$

If, on the other hand, we replace $V_0 c_{\beta_0}$ by $\Delta V + Vc_\beta$, then the identity will yield

$$s_\beta = \frac{V_0 \sin \frac{\pi}{2} \Delta i}{\Delta V} \qquad (14.48)$$

Now these expressions will readily yield

$$\tan \beta_0 = \frac{V \sin \frac{\pi}{2} \Delta i}{V_0 - V \cos \frac{\pi}{2} \Delta i} \qquad (14.49)$$

$$\tan \beta = \frac{V_0 \sin \frac{\pi}{2} \Delta i}{V_0 \cos \frac{\pi}{2} \Delta i - V} \qquad (14.50)$$

These last two expressions can be used to obtain the initial β_0 and current β provided that the appropriate Δi expression of Eqs. (14.36) or (14.37) is used according to whether $\Delta V - V_0 c_{\beta_0}$ is positive or negative. Although Eqs. (14.49) and (14.50) are valid for any optimal $(V, \Delta i)$ pair during the transfer, there is clearly a singularity at time 0 when $V = V_0$ and $\Delta i = 0$. The angle β_0 is best obtained by setting $V = V_f$ and $\Delta i = \Delta i_f$ in Eq. (14.49). It is better to use Eq. (14.41) for the control time history instead of Eq. (14.50) since we do not have to switch between two Δi expressions to describe that evolution in the first case. Finally, Edelbaum's ΔV equation in terms of the velocities and inclination is obtained from the velocity equation

$$V = \left(V_0^2 - 2V_0 \Delta V c_{\beta_0} + \Delta V^2 \right)^{1/2}$$

If we square this expression, replace ΔV in the product term by $V_0 c_{\beta_0} - V c_\beta$, and then use the identity

$$c_\beta c_{\beta_0} = \frac{1}{2} c_{\beta - \beta_0} + \frac{1}{2} c_{\beta + \beta_0}$$

with $c_{\beta - \beta_0} = \cos \pi / 2 \Delta i$ from Eq. (14.42), then we get, with $V s_\beta = V_0 s_{\beta_0}$,

$$V^2 = -V_0^2 + V_0^2 s_{\beta_0}^2 + V V_0 \cos \frac{\pi}{2} \Delta i + V_0 c_{\beta_0} V c_\beta + \Delta V^2$$

However $V_0^2 s_{\beta_0}^2 = V_0 s_{\beta_0} V s_\beta$ and, if this term is combined with $V_0 c_{\beta_0} V c_\beta$, the result will be $V V_0 c_{\beta - \beta_0}$, which can be replaced by $V V_0 \cos \pi / 2 \Delta i$. The final result is given by

$$V^2 = -V_0^2 + 2 V V_0 \cos \frac{\pi}{2} \Delta i + \Delta V^2$$

from which

$$\Delta V = \left(V_0^2 - 2 V V_0 \cos \frac{\pi}{2} \Delta i + V^2 \right)^{1/2} \tag{14.51}$$

This is Edelbaum's ΔV equation for constant-acceleration circle to inclined circle transfer. It is valid for any $(V, \Delta i)$ pair along the transfer, provided once again that the appropriate Δi expression is used, i.e., Eqs. (14.36) or (14.37) according to whether $\Delta V - V_0 c_{\beta_0}$ is <0 or >0, respectively. As shown earlier, ΔV is double-valued in the velocity since Δi itself is double-valued in that same variable. However, Eq. (14.51) is mainly used to obtain the total ΔV_{tot} required to achieve a given transfer between V_0 and V_f with a relative inclination change of Δi_f. It is valid for any $0 < \Delta i_f < 114.6$ deg or $0 < \Delta i_f < 2$ rad since this is the limiting Δi in Eq. (14.37). The transfer time t_f is simply obtained from

$$t_f = \frac{\Delta V_{\text{tot}}}{f} \tag{14.52}$$

Formulation Using Optimal Control Theory

Let the system equations be given by Eqs. (14.14) and (14.15), with time as the independent variable and i and V as the state variables. The yaw angle β is the control variable.

$$\frac{di}{dt} = \frac{2}{\pi} \frac{f}{V} s_\beta \tag{14.53}$$

$$\frac{dV}{dt} = -f c_\beta \tag{14.54}$$

This problem is now cast as a minimum time transfer problem between initial and final parameters i_0, V_0 and i_f, V_f, respectively. The variational Hamiltonian is then given by

$$H = 1 + \lambda_i \left(\frac{2}{\pi} \frac{f}{V} s_\beta \right) + \lambda_V (-f c_\beta) \qquad (14.55)$$

since the performance index is simply given by

$$J = \int_{t_0}^{t_f} L \, dt$$

with $L = 1$. The Euler–Lagrange differential equations are given by

$$\dot{\lambda}_V = -\frac{\partial H}{\partial V} = \frac{2}{\pi} \frac{f s_\beta}{V^2} \lambda_i \qquad (14.56)$$

$$\dot{\lambda}_i = -\frac{\partial H}{\partial i} = 0 \qquad (14.57)$$

Therefore, λ_i is a constant. The optimality condition is given by

$$\frac{\partial H}{\partial \beta} = \lambda_i \frac{2}{\pi} \frac{f}{V} c_\beta + f \lambda_V s_\beta = 0 \qquad (14.58)$$

which yields the optimal control law

$$\tan \beta = -\frac{2}{\pi} \frac{\lambda_i}{V \lambda_V} \qquad (14.59)$$

There is no need to integrate $\dot{\lambda}_V$ since we can use the transversality condition $H_f = 0$ at the final time. The Hamiltonian is a constant of the motion since it is not an explicit function of time. Therefore, it is equal to zero all the time. We can therefore solve for λ_V and λ_i from

$$H = 0 = 1 + \frac{2}{\pi} \frac{f}{V} s_\beta \lambda_i - f c_\beta \lambda_V$$

$$\frac{\partial H}{\partial \beta} = 0 = \frac{2}{\pi} \frac{f}{V} c_\beta \lambda_i + f s_\beta \lambda_V$$

This results in

$$\lambda_i = -\frac{\pi s_\beta V}{2f} = \text{const} \qquad (14.60)$$

$$\lambda_V = \frac{c_\beta}{f} \qquad (14.61)$$

Equation (14.60) reveals that $V s_\beta = V_0 s_{\beta_0}$ since the acceleration f is assumed to be a constant. We can now take advantage of this constancy of $V s_\beta$ in order to integrate the velocity Eq. (14.54).

$$f\,dt = -\frac{dV}{c_\beta} = \frac{-dV}{\pm\left(1 - s_\beta^2\right)^{1/2}}$$

$$f\int_0^t dt = -\int_{V_0}^V \frac{V\,dV}{\pm\left(V^2 - V_0^2 s_{\beta_0}^2\right)^{1/2}} = ft = \Delta V$$

This yields, as in the previous section,

$$V^2 = V_0^2 + \Delta V^2 - 2\Delta V V_0 c_{\beta_0}$$

$$V = \left(V_0^2 + f^2 t^2 - 2 f t V_0 c_{\beta_0}\right)^{1/2} \tag{14.62}$$

We can also obtain the preceding equation without integrating dV/dt by simply writing

$$V = \frac{V_0 s_{\beta_0}}{s_\beta} = V_0 s_{\beta_0} \frac{(1 + \tan^2 \beta)}{\tan \beta} \tag{14.63}$$

However, an expression for $\tan \beta$ is needed first. If we differentiate Eq. (14.59),

$$\frac{d}{dt}(\tan \beta) = \frac{2}{\pi} \lambda_i \frac{(\dot{V}\lambda_V + V\dot{\lambda}_V)}{V^2 \lambda_V^2} \tag{14.64}$$

Replacing $\dot{V}$ and $\dot{\lambda}_V$ by Eqs. (14.54) and (14.56) and using Eq. (14.60) and (14.61) to eliminate λ_i and λ_V, the above derivative can be written as

$$\frac{d}{d\beta}(\tan \beta)\dot{\beta} = \frac{\dot{\beta}}{c_\beta^2} = \frac{f s_\beta}{V c_\beta^2}$$

which yields

$$\dot{\beta} = \frac{f s_\beta}{V} \tag{14.65}$$

Because $V s_\beta = V_0 s_{\beta_0}$, this can also be written as

$$\dot{\beta} = \frac{f s_\beta^2}{V_0 s_{\beta_0}}$$

such that

$$\int_{\beta_0}^\beta \frac{d\beta}{s_\beta^2} = \frac{f}{V_0 s_{\beta_0}} \int_0^t dt$$

$$\cot \beta_0 - \cot \beta = \frac{f t}{V_0 s_{\beta_0}}$$

and, finally, the control law

$$\tan \beta = \frac{V_0 s_{\beta_0}}{V_0 c_{\beta_0} - ft} \tag{14.66}$$

Going back to Eq. (14.63) and replacing $\tan \beta$ with the above expression results in

$$V = \left(V_0^2 + f^2 t^2 - 2 V_0 c_{\beta_0} ft \right)^{1/2} \tag{14.67}$$

Now the inclination time history can be obtained by direct integration of Eq. (14.53) by using the expression for V in Eq. (14.67),

$$\frac{di}{dt} = \frac{2}{\pi} \frac{f}{V^2} V s_\beta = \frac{2}{\pi} \frac{f}{V^2} V_0 s_{\beta_0}$$

$$\int_0^i di = \frac{2}{\pi} V_0 s_{\beta_0} f \int_0^t \frac{dt}{V_0^2 + f^2 t^2 - 2 V_0 c_{\beta_0} ft}$$

which yields

$$\Delta i = \frac{2}{\pi} \left[\tan^{-1} \left(\frac{ft - V_0 c_{\beta_0}}{V_0 s_{\beta_0}} \right) - \tan^{-1}(- \cot \beta_0) \right]$$

Since $\tan^{-1} x = -\tan^{-1}(-x)$ and $\tan^{-1}(\cot x) = \pi/2 - \tan^{-1} x$, the final result can be written as

$$\Delta i = \frac{2}{\pi} \left[\tan^{-1} \left(\frac{ft - V_0 c_{\beta_0}}{V_0 s_{\beta_0}} \right) + \frac{\pi}{2} - \beta_0 \right] \tag{14.68}$$

This formula is uniformly valid for all t unlike the formulation of the previous section, which resulted in a set of two expressions for Δi because Δi was double-valued in the velocity. This simplification is achieved because time is selected as the independent variable instead of the velocity. If we integrate Eq. (14.56) for λ_V and use Eq. (14.60) to eliminate the constant λ_i, then,

$$\dot{\lambda}_V = -\frac{V_0^2 s_{\beta_0}^2}{V^3} = -V_0^2 s_{\beta_0}^2 \left(V_0^2 + f^2 t^2 - 2 V_0 c_{\beta_0} ft \right)^{-3/2}$$

which, upon integration, yields

$$\lambda_V = \frac{V_0 c_{\beta_0} - ft}{f V} \tag{14.69}$$

with $(\lambda_V)_0 = c_{\beta_0}/f$ and, in view of Eq. (14.61),

$$c_\beta = \frac{V_0 c_{\beta_0} - ft}{V} \tag{14.70}$$

From the definition of the influence functions λ_i and λ_V, we have

$$(\lambda_V)_0 = \frac{c_{\beta_0}}{f} = \frac{\partial t_f}{\partial V_0}$$

$$\delta t_f = \frac{c_{\beta_0}}{f} \delta V_0 \qquad (14.71)$$

$$(\lambda_i)_0 = -\frac{\pi V_0 s_{\beta_0}}{2f} = \frac{\partial t_f}{\partial (\Delta i)_0}$$

$$\delta t_f = -\frac{\pi V_0 s_{\beta_0}}{2f} \delta(\Delta i)_0 \qquad (14.72)$$

Equations (14.71) and (14.72) show how the total transfer time will vary for small variations in the initial velocity and inclination. Let us obtain an expression for ΔV in terms of β_0, V_0, and Δi. From Eq. (14.68), we have

$$\frac{ft - V_0 c_{\beta_0}}{V_0 s_{\beta_0}} = \tan\left(\frac{\pi}{2}\Delta i + \beta_0 - \frac{\pi}{2}\right) = \frac{-1}{\tan\left(\frac{\pi}{2}\Delta i + \beta_0\right)}$$

and since $ft = \Delta V$

$$\Delta V = V_0 c_{\beta_0} - \frac{V_0 s_{\beta_0}}{\tan\left(\frac{\pi}{2}\Delta i + \beta_0\right)} \qquad (14.73)$$

This expression can be used to evaluate the total ΔV for the desired transfer once β_0, the initial yaw angle, is known. To obtain β_0, let us first observe that during the integration of $\dot{V}$ in Eq. (14.54), the same intermediate results shown in the previous section and resulting from that particular integration are valid here, too. They are

$$\Delta V = ft = V_0 c_{\beta_0} - V c_\beta$$

$$\Delta V = V_0 c_{\beta_0} \pm \left(V^2 - V_0^2 s_{\beta_0}^2\right)^{1/2} \qquad (14.74)$$

From Δi in Eq. (14.68), and using the control law for $\tan\beta$ given in Eq. (14.66), we get

$$\Delta i = \frac{2}{\pi}(\beta - \beta_0) \qquad (14.75)$$

We can now obtain β_0 by using the identity in Eq. (14.74) and writing Δi as

$$\Delta i = \frac{2}{\pi}\left\{\tan^{-1}\left[\frac{\pm\left(V^2 - V_0^2 s_{\beta_0}^2\right)^{1/2}}{V_0 s_{\beta_0}}\right] + \frac{\pi}{2} - \beta_0\right\}$$

Since $\tan^{-1}x = \pm\cos^{-1}[1/(x^2 + 1)^{1/2}]$ according to whether x is >0 or <0, the preceding expression for Δi can be cast as

$$\Delta i = \frac{2}{\pi}\left[\pm\cos^{-1}\left(\frac{V_0 s_{\beta_0}}{V}\right) + \frac{\pi}{2} - \beta_0\right]$$

or

$$\cos\left\{\mp\left[\frac{\pi}{2}-\left(\beta_0+\frac{\pi}{2}\Delta i\right)\right]\right\} = \frac{V_0 s_{\beta_0}}{V}$$

or, since $\cos(-x) = \cos x$,

$$\sin\left(\beta_0+\frac{\pi}{2}\Delta i\right) = \frac{V_0 s_{\beta_0}}{V}$$

which, after expansion and division by c_{β_0}, yields

$$\tan\beta_0 = \frac{\sin\frac{\pi}{2}\Delta i}{\frac{V_0}{V}-\cos\frac{\pi}{2}\Delta i} \qquad (14.76)$$

For given V_0, V_f, and $(\Delta i)_f$, β_0 can be obtained from Eq. (14.76), which then allows us to describe V and Δi, as well as β as a function of time, from Eqs. (14.67), (14.68), and (14.66), respectively. Equation (14.76) shows that, as Δi approaches the 2-rad value or $\Delta i =$114.59 deg, $\sin\pi/2\Delta i$ will approach zero so that $\beta_0 \to 0$, indicating that the initial phase of the transfer will be coplanar. The ΔV equation of Edelbaum given by Eq. (14.51) will then approach $\Delta V = V_0 + V_f$, which is the sum of the initial and final velocities. Given that V_0 represents the ΔV needed to transfer from V_0 to ∞ or escape, and V_f represents the ΔV needed to transfer from ∞ back to V_f, the transfer is initially coplanar until escape. At infinity, $V_\infty = 0$ and the inclination change is achieved at zero cost, after which the return leg to V_f is also coplanar, resulting in $\Delta V = V_0 + V_f$. This is shown from

$$\Delta V_1 = \left(V_0^2 - 2V_0 V_\infty + V_\infty^2\right)^{1/2} = V_0$$

$$\Delta V_2 = \left(V_\infty^2 - 2V_\infty V_f + V_f^2\right)^{1/2} = V_f$$

$$\Delta V = \Delta V_1 + \Delta V_2 = V_0 + V_f \qquad (14.77)$$

For any inclination larger than $\Delta i =$114.59 deg, the cost of the orbit rotation is zero and the ΔV remains stationary at $V_0 + V_f$. If the transfer is purely coplanar, i.e., $\Delta i = 0$, then

$$\Delta V = |V_0 - V_f| \qquad (14.78)$$

the difference of the boundary velocities. For given V_0 and V_f, ΔV reaches a maximum if we set

$$\frac{\partial \Delta V}{\partial \Delta i} = 0$$

where $\Delta V = (V_0^2 - 2V_f V_0 \cos\pi/2\Delta i + V_f^2)^{1/2}$. This results in $\sin\pi/2\Delta i = 0$ or $\Delta i = 114.59$ deg as discussed earlier. Therefore, Edelbaum's Eq. (14.51) is to be used for $0 \le \Delta i \le 114.59$ deg only. For $\Delta i > 114.59$ deg, Eq. (14.77) must be used instead, as the use of Eq. (14.51) in this case will yield the wrong ΔV. Furthermore, from Eq. (14.74), as Δi approaches 114.59 deg from below, ΔV must approach $V_0 + V_f$, which implies that $\beta_0 \to 0$, as shown earlier, and $\beta_f \to 180$ deg. The ΔV in Eq. (14.73) also approaches $V_0 + V_f$ since, with $\beta_0 = \varepsilon, s_{\beta_0} \sim \beta_0$, $c_{\beta_0} \sim 1, s_{\beta_f} \sim -s_{\beta_0} \sim -\beta_0$, and $\pi/2\Delta i \sim 180$ deg.

$$\Delta V \cong V_0 - \frac{V_f s_{\beta_f}}{\tan(180 + \varepsilon)} \cong V_0 + \frac{V_f \varepsilon}{\varepsilon} = V_0 + V_f$$

Algorithm of the Edelbaum Transfer Problem

In summary, we have the following algorithm. First, one computes V_0 and V_f from the knowledge of initial and final semimajor axes a_0 and a_f, $V_0 = (\mu/a_0)^{1/2}$, $V_f = (\mu/a_f)^{1/2}$. Given $\mu = 398601.3$ km^3/s^2, Earth's gravity constant; Δi, the total inclination change desired; and f, the low-thrust acceleration, one computes β_0 from Eq. (14.76) and the total ΔV_{tot} from Eq. (14.73) such that the transfer time is known from $t_f = \Delta V_{tot}/f$. The variation with time of the various variables of interest is obtained from

$$\Delta V = ft$$

$$\beta = \tan^{-1}\left(\frac{V_0 s_{\beta_0}}{V_0 c_{\beta_0} - ft}\right)$$

$$V = \left(V_0^2 - 2V_0 ft c_{\beta_0} + f^2 t^2\right)^{1/2}$$

$$\lambda_V = \frac{c_\beta}{f}$$

$$\Delta i = \frac{2}{\pi}\left[\tan^{-1}\left(\frac{ft - V_0 c_{\beta_0}}{V_0 s_{\beta_0}}\right) + \frac{\pi}{2} - \beta_0\right]$$

λ_i is, of course, constant and given by $-\pi V_0 s_{\beta_0}/(2f)$. The total $(\Delta i)_f$ is obtained from $|i_0 - i_f|$, where i_0 and i_f are the initial and final inclination, respectively. If $i_f > i_0$, the current inclination i is given by $i = i_0 + \Delta i$. If $i_f < i_0$, then $i = i_0 - \Delta i$. This is needed since we assumed $\Delta i > 0$ so that $\beta > \beta_0$, too.

An example of a constant-acceleration LEO to GEO transfer is shown in Figs. 14.1 and 14.2. The transfer is from $a_0 = 7000$ km, $i_0 = 28.5$ deg to $a_f = 42166$ km, $i_f = 0.0$ deg with $f = 3.5 \times 10^{-7}$ km/s^2. The total transfer time $t_f = 191.26259$ days with corresponding $\Delta V_{tot} = 5.78378$ km/s. The thrust yaw angle increases from its initial value $\beta_0 = 21.98$ deg to $\beta_f = 66.75$ deg as it is more efficient to rotate the orbit plane at higher altitudes. Figures 14.3 and 14.4 show the variation of the thrust yaw angle β, semimajor axis a, velocity V, and inclination i as a function of time for a low-thrust transfer between $a_0 = 7000$ km, $i_0 = 90.0$ deg and $a_f = 42166$ km, $i_f = 0.0$ deg orbit. The acceleration is still constant at $f = 3.5 \times 10^{-7}$ km/s^2. Starting from its initial value at $\beta_0 = 10.92$ deg, β stays almost stationary for the first 100 days before surging to the final value $\beta_f = 152.29$ deg. It goes through 90 deg at day 245, where it starts to decelerate the vehicle since the orbit radius has exceeded the desired final altitude and, thus, must be shrunk until the final transfer time of $t_f = 335$ days for a total $\Delta V_{tot} = 10.13$ km/s. This is shown in Fig. 14.4, where the intermediate velocity is much less than the final desired velocity of 3.07 km/s at GEO.

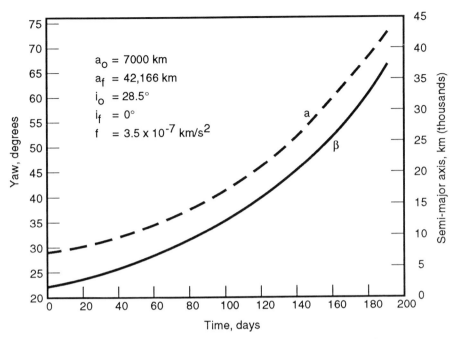

Fig. 14.1   Optimal thrust yaw profile and semimajor axis variation for a low-acceleration LEO to GEO transfer (from Ref. 1).

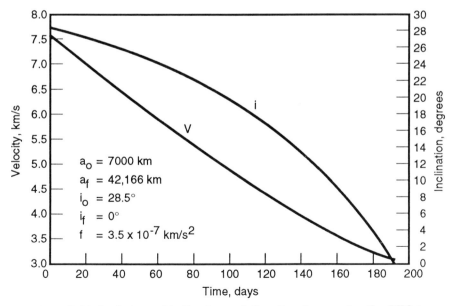

Fig. 14.2   Orbital velocity and inclination variations for a low-acceleration LEO to GEO transfer (from Ref. 1).

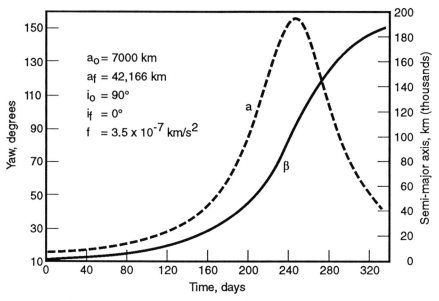

Fig. 14.3   Optimal thrust yaw profile and semimajor axis variation for a large inclination change transfer (from Ref. 1).

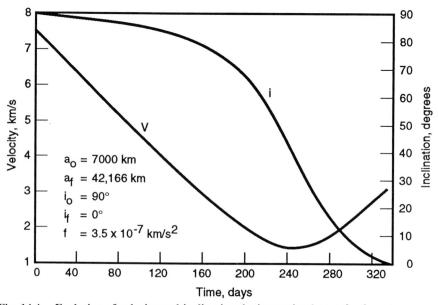

Fig. 14.4   Evolution of velocity and inclination during optimal transfer for a large inclination change case (from Ref. 1).

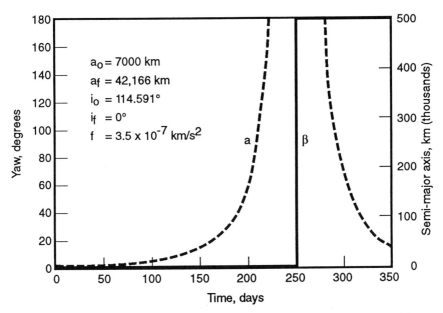

Fig. 14.5   Optimal thrust yaw profile and semimajor axis variation for the limiting case of Edelbaum's theory (Ref. 1).

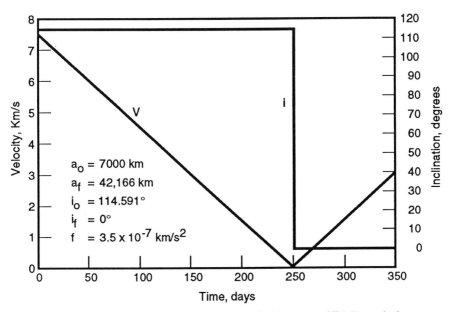

Fig. 14.6   Velocity and inclination profiles for the limiting case of Edelbaum's theory (Ref. 1).

Figures 14.5 and 14.6 are for $i_0 = 114.591$ deg, the uppermost limit of Edelbaum's
theory. The β angle stays at zero, indicating a coplanar transfer up to $t = 250$ days,
where $V \cong 0$, and the semimajor axis a at infinity, before flipping to 180 deg for
the return leg to GEO. The inclination change is carried out instantaneously at
infinity with zero cost. The total transfer time is $t_f = 351$ days with a ΔV of 10.61
km/s. This algorithm is valid regardless of whether the transfer is to a higher orbit
or a lower orbit, irrespective of the direction of the inclination change.

14.3 The Full Six-State Formulation Using Nonsingular Equinoctial Orbit Elements

The mathematical theory of orbital mechanics in terms of the nonsingular
equinoctial orbit elements has benefited from the contribution of Broucke and
Cefola in Ref. 2, who took advantage of the well-established results concerning
the Lagrange and Poisson brackets of classical elements that are found, for example, in Refs. 3 and 4 in order to transform these brackets in terms of the nonsingular
elements. Cefola later developed the single-averaged variation of parameters equations for these elements in Ref. 5, which were applied by Edelbaum, Sackett, and
Malchow in Ref. 6 to the problem of optimal low-thrust transfer. Further applications of nonsingular orbit prediction and orbit-transfer optimization problems
appeared in Refs. 7 and 8. The variation of parameters perturbation equations
based on the nonsingular equinoctial orbit elements are free from singularities for
zero eccentricity and 0- and 90-deg inclination orbits. This fact, as well as many
additional properties of the equinoctial elements, are derived in a systematic way
in Ref. 2. The matrizant, or state transition matrix corresponding to these elements,
is based on the partial derivatives of the position and velocity vectors with respect
to the equinoctial elements as well as the inverse partial derivatives, meaning the
partial derivatives of the equinoctial elements with respect to the position and
velocity vectors. These partials were derived in Ref. 2 in terms of the classical elements, and transformed later in Refs. 5 and 7 in terms of the equinoctial elements.
The applications of these elements to general and special perturbations are also
discussed in the above-mentioned references. The position and velocity compo-
nents in the direct equinoctial frame $\hat{f}$, $\hat{g}$, $\hat{w}$ of Fig. 14.7, to be described later, are
given in terms of the equinoctial elements and the eccentric longitude F, which
is related to the mean longitude λ by way of Kepler's equation. The equinoctial
elements being defined in terms of the classical elements $a, e, i, \Omega, \omega, M$, Refs. 5
and 7 consider the eccentric longitude F to be a function of λ, h, and k, where h
and k are the eccentricity vector components along $\hat{f}$ and $\hat{g}$, such that the partials
of the position and velocity vectors with respect to the equinoctial elements are
derived accordingly; F being defined by $F = E + \omega + \Omega$, where E is the classical eccentric anomaly. Given that the partials of the elements with respect to the
velocity are obtained from the partials of the position with respect to the elements
as well as the Poisson brackets,

$$\frac{\partial a_\alpha}{\partial \dot{r}} = -\sum_{\beta=1}^{6}(a_\alpha, a_\beta)\frac{\partial r}{\partial a_\beta}$$

where a_α and a_β are the generic elements and (a_α, a_β) are the Poisson brackets
of the equinoctial elements, then the partials $\partial a_\alpha/\partial \dot{r}$ are obtained in a form that

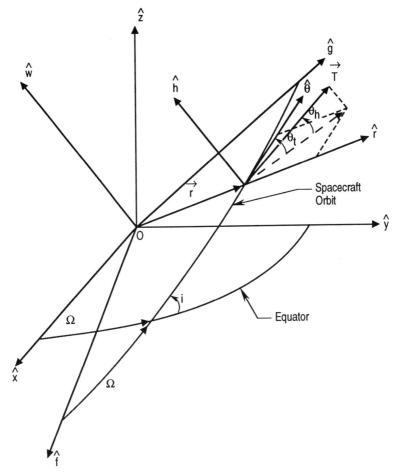

Fig. 14.7   Equinoctial frame and thrust geometry (from Ref. 15).

accounts for the dependence of F on λ, h, and k. It will be shown later how the preceding equation is used to integrate the state differential equations forward in time. In problems of optimal transfer and rendezvous, the Euler–Lagrange differential equations, which must be integrated simultaneously with the state differential equations for the solution of the two-point boundary-value problem, involve the partial derivatives of $\partial a_\alpha/\partial \dot{r}$ with respect to the equinoctial elements. The velocity partials for the first five elements were derived for the case of the optimal transfer problem in Ref. 6, by assuming that F is an independent orbit element. However, we will assume in what follows that F is dependent on λ, h, and k. This assumption is consistent with Broucke and Cefola's such that we can provide the full set of the governing equations used in the solution of space-flight optimization problems that are consistent with the original nonsingular state differential equations used in orbit prediction.

Orbital Mechanics in Terms of Equinoctial Elements

The equinoctial orbit elements were introduced in order to avoid the singularities associated with the use of the classical elements a, e, i, Ω, ω, and M. They are defined with respect to these elements by

$$a = a \qquad (14.79)$$

$$h = e \sin(\omega + \Omega) \qquad (14.80)$$

$$k = e \cos(\omega + \Omega) \qquad (14.81)$$

$$p = \tan\left(\frac{i}{2}\right) \sin \Omega \qquad (14.82)$$

$$q = \tan\left(\frac{i}{2}\right) \cos \Omega \qquad (14.83)$$

$$\lambda = M + \omega + \Omega \qquad (14.84)$$

The mean longitude λ defined in terms of the mean anomaly M can be replaced by the eccentric longitude F or the true longitude L defined respectively in terms of the eccentric anomaly E and the true anomaly θ^* such that

$$F = E + \omega + \Omega \qquad (14.85)$$

$$L = \theta^* + \omega + \Omega \qquad (14.86)$$

The inverse transformation is readily obtained from the preceding definitions:

$$a = a \qquad (14.87)$$

$$e = (h^2 + k^2)^{1/2} \qquad (14.88)$$

$$i = 2\tan^{-1}(p^2 + q^2)^{1/2} \qquad (14.89)$$

$$\Omega = \tan^{-1}\left(\frac{p}{q}\right) \qquad (14.90)$$

$$\omega = \tan^{-1}\left(\frac{h}{k}\right) - \tan^{-1}\left(\frac{p}{q}\right) \qquad (14.91)$$

$$M = \lambda - \tan^{-1}\left(\frac{h}{k}\right) \qquad (14.92)$$

The last expression can be replaced by either of the two following ones:

$$E = F - \tan^{-1}\left(\frac{h}{k}\right) \qquad (14.93)$$

$$\theta^* = L - \tan^{-1}\left(\frac{h}{k}\right) \qquad (14.94)$$

This set of equinoctial elements is called the direct equinoctial elements. They do not exhibit any singularity for $e = 0$ and $i = 0$ deg, 90 deg. However, they are singular for $i = 180$ deg. The equinoctial frame $\hat{f}$, $\hat{g}$, $\hat{w}$ is defined in Fig.14.7 respective to the inertial frame $\hat{x}$, $\hat{y}$, $\hat{z}$. The direction of the unit vector $\hat{f}$ is such that both $\hat{f}$ and $\hat{g}$ are in the orbit plane and $\hat{f}$ is obtained through a clockwise rotation of an angle Ω from the direction of the ascending node. The three rotations Ω, $-i$, $-\Omega$ transform the $\hat{f}$, $\hat{g}$, $\hat{w}$ system into the $\hat{x}$, $\hat{y}$, $\hat{z}$ system such that

$$
\begin{pmatrix} \hat{x} \\ \hat{y} \\ \hat{z} \end{pmatrix} = \begin{pmatrix} c_\Omega c_{-\Omega} - s_\Omega c_i s_{-\Omega} & -c_\Omega s_{-\Omega} - s_\Omega c_i c_{-\Omega} & s_\Omega s_i \\ s_\Omega c_{-\Omega} + c_\Omega c_i s_{-\Omega} & -s_\Omega s_{-\Omega} + c_\Omega c_i c_{-\Omega} & -c_\Omega s_i \\ s_i s_{-\Omega} & s_i c_{-\Omega} & c_i \end{pmatrix} \begin{pmatrix} \hat{f} \\ \hat{g} \\ \hat{w} \end{pmatrix} \tag{14.95}
$$

which reduces to

$$
\begin{pmatrix} \hat{x} \\ \hat{y} \\ \hat{z} \end{pmatrix} = \begin{pmatrix} c_\Omega^2 + c_i s_\Omega^2 & c_\Omega s_\Omega - s_\Omega c_i c_\Omega & s_\Omega s_i \\ s_\Omega c_\Omega - s_\Omega c_\Omega c_i & s_\Omega^2 + c_i c_\Omega^2 & -c_\Omega s_i \\ -s_\Omega s_i & s_i c_\Omega & c_i \end{pmatrix} \begin{pmatrix} \hat{f} \\ \hat{g} \\ \hat{w} \end{pmatrix} \tag{14.96}
$$

The inverse transformation is given by

$$
\begin{pmatrix} \hat{f} \\ \hat{g} \\ \hat{w} \end{pmatrix} = \begin{pmatrix} c_\Omega^2 + c_i s_\Omega^2 & s_\Omega c_\Omega - s_\Omega c_\Omega c_i & -s_\Omega s_i \\ s_\Omega c_\Omega - s_\Omega c_\Omega c_i & s_\Omega^2 + c_i c_\Omega^2 & s_i c_\Omega \\ s_\Omega s_i & -c_\Omega s_i & c_i \end{pmatrix} \begin{pmatrix} \hat{x} \\ \hat{y} \\ \hat{z} \end{pmatrix} \tag{14.97}
$$

From the definitions of the equinoctial elements, the following relationships can be established:

$$
c_{2\Omega} = \frac{q^2 - p^2}{p^2 + q^2}
$$

$$
s_\Omega = \frac{p}{(p^2 + q^2)^{1/2}}
$$

$$
c_\Omega = \frac{q}{(p^2 + q^2)^{1/2}}
$$

$$
s_{i/2}^2 = \frac{p^2 + q^2}{1 + p^2 + q^2}
$$

$$
c_{i/2}^2 = \frac{1}{1 + p^2 + q^2}
$$

$$
s_i = \frac{2(p^2 + q^2)^{1/2}}{1 + p^2 + q^2}
$$

$$c_i = \frac{1 - p^2 - q^2}{1 + p^2 + q^2}$$

$$(1 + c_i)^2 = \frac{4}{(1 + p^2 + q^2)^2}$$

It is then possible to obtain expressions relating the various elements of the transformation matrix in Eq. (14.97) in terms of the equinoctial elements p and q such that

$$c_\Omega^2 + c_i s_\Omega^2 = \frac{1 - p^2 + q^2}{1 + p^2 + q^2}$$

$$s_\Omega c_\Omega - s_\Omega c_\Omega c_i = \frac{2pq}{1 + p^2 + q^2}$$

$$s_\Omega s_i = \frac{2p}{1 + p^2 + q^2}$$

$$s_\Omega^2 + c_i c_\Omega^2 = \frac{1 + p^2 - q^2}{1 + p^2 + q^2}$$

$$s_i c_\Omega = \frac{2q}{1 + p^2 + q^2}$$

From Eq. (14.97), the components of the $\hat{f}$, $\hat{g}$, $\hat{w}$ unit vectors in the inertial $\hat{x}$, $\hat{y}$, $\hat{z}$ frame are now readily obtained in terms of the direct equinoctial elements

$$\hat{f} = \frac{1}{(1 + p^2 + q^2)} \begin{pmatrix} 1 - p^2 + q^2 \\ 2pq \\ -2p \end{pmatrix} \qquad (14.98)$$

$$\hat{g} = \frac{1}{(1 + p^2 + q^2)} \begin{pmatrix} 2pq \\ 1 + p^2 - q^2 \\ 2q \end{pmatrix} \qquad (14.99)$$

$$\hat{w} = \frac{1}{(1 + p^2 + q^2)} \begin{pmatrix} 2p \\ -2q \\ 1 - p^2 - q^2 \end{pmatrix} \qquad (14.100)$$

The derivation of Kepler's equation in terms of the equinoctial elements is carried out by using the definitions of e as well as h and k in the classical Kepler equation,

$$M = E - e \sin E$$

$$\lambda - (\omega + \Omega) = F - (\omega + \Omega) - e \sin[F - (\omega + \Omega)]$$

$$\lambda = F - e\left(s_F \frac{k}{e} - c_F \frac{h}{e}\right)$$

$$\lambda = F - k \sin F + h \cos F \tag{14.101}$$

The orbit equation $r = a(1 - e \cos E)$ can be written in terms of the equinoctial elements by using $F = E + \omega + \Omega$ and the definitions in Eqs. (14.80), (14.81), and (14.88), which relate e to h and k:

$$r = a\{1 - e \cos[F - (\omega + \Omega)]\}$$

$$r = a(1 - k \cos F - h \sin F) \tag{14.102}$$

With the radius vector $\mathbf{r}$ being contained in the orbit plane, its components X_1 and Y_1 along $\hat{f}$ and $\hat{g}$ must be derived. Then, $\dot{\mathbf{r}} = \dot{X}_1 \hat{f} + \dot{Y}_1 \hat{g}$ can also be evaluated. To do this, let us observe that $X_1 = r \cos(\theta^* + \omega + \Omega)$ and $Y_1 = r \sin(\theta^* + \omega + \Omega)$. From classical orbital mechanics, the relations $r c_{\theta^*} = a(\cos E - e)$ and $r s_{\theta^*} = a(1 - e^2)^{1/2} \sin E$ can now be used before the definitions in Eqs. (14.80), (14.81), (14.85), and (14.88) are introduced. Then,

$$X_1 = r c_{\theta^*} c_{\omega+\Omega} - r s_{\theta^*} s_{\omega+\Omega}$$

$$= a\left\{\frac{k^2 + h^2(1 - h^2 - k^2)^{1/2}}{(h^2 + k^2)} c_F + \frac{hk\left[1 - (1 - h^2 - k^2)^{1/2}\right]}{(h^2 + k^2)} s_F - k\right\}$$

This intermediary result is written in order to introduce the quantity β, which is not to be confused with the thrust yaw angle β of the Edelbaum theory described earlier.

$$\beta = \frac{1 - (1 - h^2 - k^2)^{1/2}}{(h^2 + k^2)}$$

$$\beta = \frac{1}{1 + (1 - h^2 - k^2)^{1/2}} \tag{14.103}$$

We observe that

$$\frac{k^2 + h^2(1 - h^2 - k^2)^{1/2}}{(h^2 + k^2)} = 1 - h^2\beta$$

and find that X_1 takes the final form

$$X_1 = a[(1 - h^2\beta) \cos F + hk\beta \sin F - k] \tag{14.104}$$

The same manipulations are carried out for the definition of Y_1, which reduces to

$$Y_1 = a[hk\beta \cos F + (1 - k^2\beta) \sin F - h] \tag{14.105}$$

The velocity components $\dot{X}_1$ and $\dot{Y}_1$ can be obtained directly from Eqs. (14.104) and (14.105) by holding a, h, and k, therefore, also β constant and varying only F.

$$\dot{X}_1 = a\left[-(1 - h^2\beta)s_F\dot{F} + hk\beta c_F\dot{F} \right]$$

$$\dot{Y}_1 = a\left[-hk\beta s_F\dot{F} + (1 - k^2\beta)c_F\dot{F} \right]$$

From Eq. (14.101) and with t_0 designating time at epoch,

$$\lambda = n(t - t_0) + \lambda_0 = F - ks_F + hc_F$$

$$\dot{\lambda} = n = \dot{F} - kc_F\dot{F} - hs_F\dot{F}$$

$$\dot{F} = \frac{n}{(1 - kc_F - hs_F)}$$

and, in view of Eq. (14.102), the orbit equation

$$\dot{F} = \frac{na}{r} \tag{14.106}$$

The velocity components can then be cast into the following form:

$$\dot{X}_1 = \frac{a^2 n}{r}[hk\beta c_F - (1 - h^2\beta)s_F] \tag{14.107}$$

$$\dot{Y}_1 = \frac{na^2}{r}[(1 - k^2\beta)c_F - hk\beta s_F] \tag{14.108}$$

Given r and $\dot{r}$ in the inertial $\hat{x}, \hat{y}, \hat{z}$ system, the equinoctial elements can be obtained by first evaluating a and e with the semimajor axis obtained from the energy equation

$$a = \left(\frac{2}{|r|} - \frac{|\dot{r}|^2}{\mu} \right)^{-1} \tag{14.109}$$

$$e = -\frac{r}{|r|} - \frac{(r \times \dot{r}) \times \dot{r}}{\mu} \tag{14.110}$$

Since $\hat{w}$ is along the angular momentum vector $h = r \times \dot{r}$, we have

$$\hat{w} = \frac{r \times \dot{r}}{|r \times \dot{r}|} = \begin{pmatrix} w_x \\ w_y \\ w_z \end{pmatrix} \tag{14.111}$$

From the definition of $\hat{w}$ given by Eq. (14.100), it follows that p and q can be determined from Eq. (14.111) such that

$$p = \frac{w_x}{(1 + w_z)} \qquad (14.112)$$

$$q = \frac{-w_y}{(1 + w_z)} \qquad (14.113)$$

Since k and h are the components of the eccentricity vector e along $\hat{f}$ and $\hat{g}$, we have

$$k = e \cdot \hat{f} \qquad (14.114)$$

$$h = e \cdot \hat{g} \qquad (14.115)$$

Finally, the components of r along $\hat{f}$ and $\hat{g}$ are obtained from

$$X_1 = r \cdot \hat{f} \qquad (14.116)$$

$$Y_1 = r \cdot \hat{g} \qquad (14.117)$$

These quantities are needed to evaluate the mean longitude λ, which itself requires the evaluation of the eccentric longitude F since $\lambda = F - k s_F + h c_F$. The quantities s_F and c_F are obtained by solving the system of Eqs. (14.104) and (14.105), which express X_1 and Y_1 in terms of h, k, and F. This results in

$$s_F = h + \frac{Y_1(1 - h^2\beta) - hk\beta X_1}{a(1 - h^2 - k^2)^{1/2}}$$

$$c_F = \frac{(1 - k^2\beta)X_1 - hk\beta Y_1}{a(1 - h^2 - k^2)^{1/2}} + k$$

X_1 and Y_1 are first obtained from Eqs. (14.116) and (14.117). Finally, the partials of X_1, Y_1, $\dot{X}_1$, and $\dot{Y}_1$ with respect to h and k, which will be needed later, are evaluated from Eqs. (14.104), (14.105), (14.107), and (14.108). The following partials must, however, first be derived from

$$\beta = \frac{1}{1 + (1 - h^2 - k^2)^{1/2}}$$

with the observation that $(1 - h^2 - k^2)^{1/2} = (1 - \beta)/\beta$. They are

$$\frac{\partial \beta}{\partial h} = \frac{h\beta^3}{(1 - \beta)}$$

$$\frac{\partial \beta}{\partial k} = \frac{k\beta^3}{(1 - \beta)}$$

From $\lambda = F - ks_F + hc_F = M + \omega + \Omega$, it follows that

$$\frac{\partial \lambda}{\partial h} = 0 = \frac{\partial F}{\partial h} - kc_F \frac{\partial F}{\partial h} + c_F - hs_F \frac{\partial F}{\partial h}$$

$$= \frac{\partial F}{\partial h}[1 - kc_F - hs_F] + c_F = \frac{\partial F}{\partial h}\left(\frac{r}{a}\right) + c_F$$

$$\frac{\partial F}{\partial h} = -\frac{a}{r}c_F \qquad (14.118)$$

Similarly, from $\partial \lambda / \partial k = 0$,

$$\frac{\partial F}{\partial k} = \frac{a}{r}s_F \qquad (14.119)$$

which is then used in Eq. (14.104), together with $\partial \beta / \partial h$ developed earlier, to yield, after some manipulations,

$$\frac{\partial X_1}{\partial h} = a\left[- (hc_F - ks_F)\left(\beta + \frac{h^2\beta^3}{(1-\beta)}\right) - \frac{a}{r}c_F(h\beta - s_F)\right] \qquad (14.120)$$

In a similar way, and as in Ref. 5,

$$\frac{\partial X_1}{\partial k} = -a\left[(hc_F - ks_F)\frac{hk\beta^3}{(1-\beta)} + 1 + \frac{a}{r}s_F(s_F - h\beta)\right] \qquad (14.121)$$

$$\frac{\partial Y_1}{\partial h} = a\left[(hc_F - ks_F)\frac{hk\beta^3}{(1-\beta)} - 1 + \frac{a}{r}c_F(k\beta - c_F)\right] \qquad (14.122)$$

$$\frac{\partial Y_1}{\partial k} = a\left[(hc_F - ks_F)\left(\beta + \frac{k^2\beta^3}{(1-\beta)}\right) + \frac{a}{r}s_F(c_F - k\beta)\right] \qquad (14.123)$$

Letting $z = (a\ h\ k\ p\ q\ \lambda_0)^T$ represent the orbit state vector at time t, with λ_0 the mean longitude at epoch corresponding to the mean anomaly M_0, and given that $z = f(r, \dot{r})$, we have

$$\dot{z} = \frac{\partial z}{\partial r}\dot{r} + \frac{\partial z}{\partial \dot{r}}\ddot{r} \qquad (14.124)$$

where

$$\ddot{r} = \frac{T}{m} + g = \frac{T}{m} - \frac{\mu}{r^3}r \qquad (14.125)$$

and where T and m stand for the thrust vector and spacecraft mass, respectively. The partial derivatives of the equinoctial elements with respect to the position and velocity vectors, the so-called inverse partials, are related to the partials of the position and velocity vectors with respect to the equinoctial elements using the Poisson brackets (a_α, a_β) of equinoctial elements.

$$\frac{\partial a_\alpha}{\partial r} = \sum_{\beta=1}^{6}(a_\alpha, a_\beta)\frac{\partial \dot{r}}{\partial a_\beta}$$ (14.126)

$$\frac{\partial a_\alpha}{\partial \dot{r}} = -\sum_{\beta=1}^{6}(a_\alpha, a_\beta)\frac{\partial r}{\partial a_\beta}$$ (14.127)

If we write Eq. (14.124) one component at a time, then, using Eq. (14.125),

$$\dot{z} = \frac{\partial z}{\partial r}\dot{r} + \frac{\partial z}{\partial \dot{r}}\ddot{r}$$

$$= \sum_\beta (a_\alpha, a_\beta)\frac{\partial \dot{r}}{\partial a_\beta}\dot{r} - \sum_\beta(a_\alpha, a_\beta)\frac{\partial r}{\partial a_\beta}\ddot{r}$$

$$= \sum_\beta (a_\alpha, a_\beta)\frac{\partial \dot{r}}{\partial a_\beta}\dot{r} + \sum_\beta(a_\alpha, a_\beta)\frac{\partial r}{\partial a_\beta}\left(\frac{\mu}{r^3}\right)r + \frac{\partial z}{\partial \dot{r}}\frac{T}{m}$$

$$= \sum_\beta (a_\alpha, a_\beta)\left(\frac{\partial \dot{r}}{\partial a_\beta}\dot{r} + \frac{\mu}{r^3}\frac{\partial r}{\partial a_\beta}r\right) + \frac{\partial z}{\partial \dot{r}}\frac{T}{m}$$ (14.128)

The term in brackets can be written as

$$\frac{1}{2}\frac{\partial(\dot{r}^2)}{\partial z} + \frac{\mu}{r^3}\frac{1}{2}\frac{\partial(r^2)}{\partial z} = \frac{1}{2}\left(\frac{\partial v^2}{\partial z} + \frac{\mu}{r^3}\frac{\partial r^2}{\partial z}\right)$$ (14.129)

Since, $|\dot{r}| = v$ and, from the energy equation,

$$\frac{1}{2}v^2 - \frac{\mu}{r} = C$$

we have

$$\frac{1}{2}\frac{\partial v^2}{\partial z} = -\frac{\mu}{r^2}\frac{\partial r}{\partial z}$$

and, in view of $\partial r^2/\partial z = 2r\,\partial r/\partial z$, the bracket in Eq. (14.129) cancels out such that

$$\dot{z} = \frac{\partial z}{\partial \dot{r}}\frac{T}{m}$$

If we let $\hat{u}$ represent a unit vector in the direction of the thrust,

$$\boldsymbol{T} = T\hat{u}$$ (14.130)

and write z as a vector, we obtain the variation of parameters equations, where both $\partial z / \partial r$ and $\hat{u}$ are expressed in the direct equinoctial frame,

$$\dot{z} = \frac{\partial z}{\partial \dot{r}} \frac{T}{m} \hat{u} = \frac{\partial z}{\partial \dot{r}} f \hat{u} \qquad (14.131)$$

Partials of the Equinoctial Orbit Elements with Respect to Velocity

The first-order differential equations in the equinoctial orbit elements involve the partials of these elements with respect to the velocity vector $\dot{r}$. These partials are computed using the Poisson brackets of the equinoctial elements (a_α, a_β). However, the partials of the position vector r with respect to the equinoctial elements must be computed first. Furthermore, the Poisson brackets of the equinoctial elements can be obtained from the Poisson brackets of the classical elements, which are found in texts such as Refs. 3 and 4. These brackets have been derived in Refs. 2 and 5. These "velocity partials" are given by Eq. (14.127). First, let us derive the $\partial r / \partial a_\beta$ partials. From the definition of $r = X_1 \hat{f} + Y_1 \hat{g}$, it follows that

$$\frac{\partial r}{\partial a} = \frac{\partial X_1}{\partial a} \hat{f} + \frac{\partial Y_1}{\partial a} \hat{g}$$

Refer back to the definition of λ to arrive at

$$\frac{\partial \lambda}{\partial a} = \frac{\partial \lambda}{\partial F} \frac{\partial F}{\partial a} = -\frac{3}{2a} nt$$

$$\frac{\partial F}{\partial a} = -\frac{3}{2} \frac{n}{r} t \qquad (14.132)$$

Then, use the definitions of $\dot{X}_1$ and $\dot{Y}_1$ in Eqs. (14.107) and (14.108) to obtain

$$\frac{\partial X_1}{\partial a} = [(1 - h^2 \beta)c_F + hk\beta s_F - k] + a[-(1 - h^2\beta)s_F + hk\beta c_F] \frac{\partial F}{\partial a}$$

$$= \frac{X_1}{a} + \frac{r \dot{X}_1}{na} \frac{\partial F}{\partial a} = \frac{X_1}{a} - \frac{3}{2} \frac{t}{a} \dot{X}_1$$

In a similar way,

$$\frac{\partial Y_1}{\partial a} = \frac{Y_1}{a} - \frac{3}{2} \frac{t}{a} \dot{Y}_1$$

and, therefore,

$$\frac{\partial r}{\partial a} = \left(\frac{X_1}{a} - \frac{3}{2} \frac{t}{a} \dot{X}_1 \right) \hat{f} + \left(\frac{Y_1}{a} - \frac{3}{2} \frac{t}{a} \dot{Y}_1 \right) \hat{g}$$

The partials of $\partial r / \partial h$ and $\partial r / \partial k$ are written in a straightforward manner:

$$\frac{\partial r}{\partial h} = \frac{\partial X_1}{\partial h} \hat{f} + \frac{\partial Y_1}{\partial h} \hat{g} \qquad (14.133)$$

$$\frac{\partial r}{\partial k} = \frac{\partial X_1}{\partial k} \hat{f} + \frac{\partial Y_1}{\partial k} \hat{g} \qquad (14.134)$$

Equations (14.120–14.123) must, of course, be used in Eqs. (14.133) and (14.134). Now, by way of some manipulations,

$$\frac{\partial \boldsymbol{r}}{\partial \lambda} = \frac{\partial \boldsymbol{r}}{\partial F}\frac{\partial F}{\partial \lambda} = \left(\frac{\partial X_1}{\partial F}\hat{f} + \frac{\partial Y_1}{\partial F}\hat{g}\right)\frac{\partial F}{\partial \lambda}$$

$$\frac{\partial \boldsymbol{r}}{\partial \lambda} = \frac{\dot{X}_1}{n}\hat{f} + \frac{\dot{Y}_1}{n}\hat{g} = \frac{\dot{\boldsymbol{r}}}{n} \tag{14.135}$$

The remaining partials $\partial \boldsymbol{r}/\partial p$ and $\partial \boldsymbol{r}/\partial q$ are derived by observing first that X_1 and Y_1 are not functions of either p or q but that the components of the unit vectors $\hat{f}$ and $\hat{g}$ are such functions; therefore,

$$\frac{\partial \boldsymbol{r}}{\partial p} = X_1\frac{\partial \hat{f}}{\partial p} + Y_1\frac{\partial \hat{g}}{\partial p}$$

From Eqs. (14.98–14.100), it follows that

$$\frac{\partial \boldsymbol{r}}{\partial p} = \frac{2X_1}{(1 + p^2 + q^2)^2}\begin{pmatrix} -2p - 2pq^2 \\ 2q - q(1 + p^2 - q^2) \\ -2q^2 - (1 - p^2 - q^2) \end{pmatrix}$$

$$+ \frac{2Y_1}{(1 + p^2 + q^2)^2}\begin{pmatrix} q(1 + q^2 - p^2) \\ 2pq^2 \\ -2pq \end{pmatrix}$$

$$\frac{\partial \boldsymbol{r}}{\partial p} = \frac{2}{(1 + p^2 + q^2)}[q(Y_1\hat{f} - X_1\hat{g}) - X_1\hat{w}] \tag{14.136}$$

In a similar way,

$$\frac{\partial \boldsymbol{r}}{\partial q} = X_1\frac{\partial \hat{f}}{\partial q} + Y_1\frac{\partial \hat{g}}{\partial q} = \frac{2}{(1 + p^2 + q^2)}[p(X_1\hat{g} - Y_1\hat{f}) + Y_1\hat{w}] \tag{14.137}$$

The Poisson brackets of equinoctial elements are very tedious to derive unless one uses the transformation equation that converts the Poisson brackets of classical elements into the brackets of equinoctial elements. The transformation equation is given in Ref. 2 and, using the nomenclature introduced by Broucke and Cefola, we write

$$[(p_\alpha, p_\beta)] = \left[\frac{\partial p_\alpha}{\partial a_\lambda}\right][(a_\lambda, a_\mu)]\left[\frac{\partial p_\beta}{\partial a_\mu}\right]^T \tag{14.138}$$

The various matrices appearing in Eq. (14.138) are given by

$$[(p_\alpha, p_\beta)] = \begin{bmatrix} (a,a) & (a,h) & (a,k) & (a,\lambda_0) & (a,p) & (a,q) \\ (h,a) & (h,h) & (h,k) & (h,\lambda_0) & (h,p) & (h,q) \\ (k,a) & (k,h) & (k,k) & (k,\lambda_0) & (k,p) & (k,q) \\ (\lambda_0,a) & (\lambda_0,h) & (\lambda_0,k) & (\lambda_0,\lambda_0) & (\lambda_0,p) & (\lambda_0,q) \\ (p,a) & (p,h) & (p,k) & (p,\lambda_0) & (p,p) & (p,q) \\ (q,a) & (q,h) & (q,k) & (q,\lambda_0) & (q,p) & (q,q) \end{bmatrix}$$

$$[(a_\lambda, a_\mu)] = \begin{bmatrix} (a,a) & (a,e) & (a,i) & (a,\Omega) & (a,\omega) & (a,M_0) \\ (e,a) & (e,e) & (e,i) & (e,\Omega) & (e,\omega) & (e,M_0) \\ (i,a) & (i,e) & (i,i) & (i,\Omega) & (i,\omega) & (i,M_0) \\ (\Omega,a) & (\Omega,e) & (\Omega,i) & (\Omega,\Omega) & (\Omega,\omega) & (\Omega,M_0) \\ (\omega,a) & (\omega,e) & (\omega,i) & (\omega,\Omega) & (\omega,\omega) & (\omega,M_0) \\ (M_0,a) & (M_0,e) & (M_0,i) & (M_0,\Omega) & (M_0,\omega) & (M_0,M_0) \end{bmatrix}$$

$$\left[\frac{\partial p_\alpha}{\partial a_\lambda} \right] = \begin{bmatrix} \dfrac{\partial a}{\partial a} & \dfrac{\partial a}{\partial e} & \dfrac{\partial a}{\partial i} & \dfrac{\partial a}{\partial \Omega} & \dfrac{\partial a}{\partial \omega} & \dfrac{\partial a}{\partial M_0} \\[2mm] \dfrac{\partial h}{\partial a} & \dfrac{\partial h}{\partial e} & \dfrac{\partial h}{\partial i} & \dfrac{\partial h}{\partial \Omega} & \dfrac{\partial h}{\partial \omega} & \dfrac{\partial h}{\partial M_0} \\[2mm] \dfrac{\partial k}{\partial a} & \dfrac{\partial k}{\partial e} & \dfrac{\partial k}{\partial i} & \dfrac{\partial k}{\partial \Omega} & \dfrac{\partial k}{\partial \omega} & \dfrac{\partial k}{\partial M_0} \\[2mm] \dfrac{\partial \lambda_0}{\partial a} & \dfrac{\partial \lambda_0}{\partial e} & \dfrac{\partial \lambda_0}{\partial i} & \dfrac{\partial \lambda_0}{\partial \Omega} & \dfrac{\partial \lambda_0}{\partial \omega} & \dfrac{\partial \lambda_0}{\partial M_0} \\[2mm] \dfrac{\partial p}{\partial a} & \dfrac{\partial p}{\partial e} & \dfrac{\partial p}{\partial i} & \dfrac{\partial p}{\partial \Omega} & \dfrac{\partial p}{\partial \omega} & \dfrac{\partial p}{\partial M_0} \\[2mm] \dfrac{\partial q}{\partial a} & \dfrac{\partial q}{\partial e} & \dfrac{\partial q}{\partial i} & \dfrac{\partial q}{\partial \Omega} & \dfrac{\partial q}{\partial \omega} & \dfrac{\partial q}{\partial M_0} \end{bmatrix}$$

$$\left[\frac{\partial p_\beta}{\partial a_\mu} \right]^T = \begin{bmatrix} \dfrac{\partial a}{\partial a} & \dfrac{\partial h}{\partial a} & \dfrac{\partial k}{\partial a} & \dfrac{\partial \lambda_0}{\partial a} & \dfrac{\partial p}{\partial a} & \dfrac{\partial q}{\partial a} \\[2mm] \dfrac{\partial a}{\partial e} & \dfrac{\partial h}{\partial e} & \dfrac{\partial k}{\partial e} & \dfrac{\partial \lambda_0}{\partial e} & \dfrac{\partial p}{\partial e} & \dfrac{\partial q}{\partial e} \\[2mm] \dfrac{\partial a}{\partial i} & \dfrac{\partial h}{\partial i} & \dfrac{\partial k}{\partial i} & \dfrac{\partial \lambda_0}{\partial i} & \dfrac{\partial p}{\partial i} & \dfrac{\partial q}{\partial i} \\[2mm] \dfrac{\partial a}{\partial \Omega} & \dfrac{\partial h}{\partial \Omega} & \dfrac{\partial k}{\partial \Omega} & \dfrac{\partial \lambda_0}{\partial \Omega} & \dfrac{\partial p}{\partial \Omega} & \dfrac{\partial q}{\partial \Omega} \\[2mm] \dfrac{\partial a}{\partial \omega} & \dfrac{\partial h}{\partial \omega} & \dfrac{\partial k}{\partial \omega} & \dfrac{\partial \lambda_0}{\partial \omega} & \dfrac{\partial p}{\partial \omega} & \dfrac{\partial q}{\partial \omega} \\[2mm] \dfrac{\partial a}{\partial M_0} & \dfrac{\partial h}{\partial M_0} & \dfrac{\partial k}{\partial M_0} & \dfrac{\partial \lambda_0}{\partial M_0} & \dfrac{\partial p}{\partial M_0} & \dfrac{\partial q}{\partial M_0} \end{bmatrix}$$

for the numerical integration of the $\dot{\lambda}_z$ rates. The only nonzero $\partial n/\partial z$ partial is given by

$$\frac{\partial n}{\partial a} = -\frac{3n}{2a} = -\frac{3}{2}\mu^{1/2}a^{-5/2}$$

The following partials are used to generate the $\partial M/\partial z$ partials of the Appendix. Here F must be considered to be independent of a but not of h, k, and λ because λ is being integrated and is therefore independent of a, h, k, p, and q such that $\partial\lambda/\partial a$ is now equal to zero. From Kepler's equation, this implies that $\partial F/\partial a = 0$. Thus,

$$\frac{\partial r}{\partial a} = \frac{r}{a}$$

$$\frac{\partial r}{\partial h} = \frac{a^2}{r}(h - s_F)$$

$$\frac{\partial r}{\partial k} = \frac{a^2}{r}(k - c_F)$$

$$\frac{\partial r}{\partial F} = a(ks_F - hc_F)$$

$$\frac{\partial F}{\partial a} = 0$$

$$\frac{\partial F}{\partial h} = -\frac{a}{r}c_F$$

$$\frac{\partial F}{\partial k} = \frac{a}{r}s_F$$

$$\frac{\partial F}{\partial \lambda} = \frac{a}{r}$$

It is also true that $\partial\lambda/\partial F = r/a$. The components u_f, u_g, u_w of the unit vector $\hat{u}$ in the equinoctial frame are obtained from

$$\hat{u} = \left(\boldsymbol{\lambda}_z^T M\right)^T / \left|\boldsymbol{\lambda}_z^T M\right| \tag{14.156}$$

From Fig. 14.7, the thrust vector $\boldsymbol{T} = T\hat{u}$ is defined by the thrust pitch and thrust yaw angles θ_t and θ_h, respectively, in the rotating $\hat{r}$, $\hat{\theta}$, $\hat{h}$ frame; $\hat{r}$ is a unit vector along the instantaneous position vector $\boldsymbol{r}$, with $\hat{\theta}$ in the orbit plane and along the direction of motion, and $\hat{h}$ along the angular momentum vector. Therefore,

$$\theta_t = \tan^{-1}(u_r/u_\theta) \tag{14.157}$$

$$\theta_h = \tan^{-1}(u_w/u_\theta) \tag{14.158}$$

with u_r, u_θ, and u_h, representing the components of the unit vector $\hat{u}$ along the $\hat{r}$, $\hat{\theta}$, and $\hat{h}$ directions. We have

$$\hat{r} = \frac{X_1}{|r|}\hat{f} + \frac{Y_1}{|r|}\hat{g} \tag{14.159}$$

$$\hat{\theta} = \hat{h} \times \hat{r} = -\frac{Y_1}{|r|}\hat{f} + \frac{X_1}{|r|}\hat{g} \tag{14.160}$$

such that

$$u_r = \frac{X_1}{r}u_f + \frac{Y_1}{r}u_g \tag{14.161}$$

$$u_\theta = -\frac{Y_1}{r}u_f + \frac{X_1}{r}u_g \tag{14.162}$$

$$u_h = u_w \tag{14.163}$$

where

$$r = |r| = \left(X_1^2 + Y_1^2\right)^{1/2}$$

14.4 Orbit Transfer with Continuous Constant Acceleration

This section presents examples of precision-integrated, optimized low earth orbit (LEO) to geostationary earth orbit (GEO) minimum-time transfer and compares them to the solutions obtained by way of the averaging technique. A 10^{-2} g acceleration applied in a constant and continuous manner is taken as an example in order to generate fast subday transfers that could be flown with nuclear thermal propulsion upper stages. The six-state formulation used here allows the user to generate optimal transfers that first start from a given fixed location on the initial orbit while optimizing the arrival point on the target or final orbit. The analysis is further extended to optimize both departure and arrival points in order to obtain the overall minimum-time free-free solution. This requires the vanishing of the Lagrange multiplier adjoint to the mean longitude at both initial and final times with fixed initial time and optimized final time. These fast, few-revolution, five-state transfers are sensitive to initial and final orbital position, thereby necessitating the use of the full six-state dynamics. These exact results are then compared to the approximate solutions obtained using averaged dynamics with robust and fast convergence characteristics. These examples determine that the ΔVs or transfer time solutions compare rather well, even for these short-duration transfers, but that the element time histories, and especially the eccentricity, are poorly simulated by the approximate solutions. Furthermore, due to the nature of the averaging technique, the sensitivity of the solution to orbital position is totally removed such that the precision-integrated solution must be used instead for accurate guidance.

Minimum-Time Transfer from Fixed Initial State with Continuous Constant Acceleration

We minimize total transfer time by maximizing the performance index $J = \int_{t_0}^{t_f} L \, dt = -\int_{t_0}^{t_f} dt = -(t_f - t_0)$. For fixed t_0, the minimization of t_f or maximization of $-t_f$ gives rise to the transversality condition $H_f = 1$ for $H = \lambda_z^T \dot{z}$ since $H_f = 0$ for the augmented Hamiltonian $H = -1 + \lambda_z^T \dot{z}$. We now use the full six-element formulation to solve five-state orbit-transfer problems by first starting from a fixed initial state and optimizing the final arrival point on the terminal orbit. Given $(a)_0$, $(e)_0$, $(i)_0$, $(\Omega)_0$, $(\omega)_0$, and $(M)_0$ at time $t_0 = 0$ or, equivalently, $(a)_0$, $(h)_0$, $(k)_0$, $(p)_0$, $(q)_0$, and $(\lambda)_0$, the initial values of the Lagrange multipliers, namely, $(\lambda_a)_0$, $(\lambda_h)_0$, $(\lambda_k)_0$, $(\lambda_p)_0$, $(\lambda_q)_0$, and $(\lambda_\lambda)_0$ are guessed, and the dynamic equations $\dot{z} = (\partial z / \partial r) \cdot \hat{u} f$ as well as the adjoint Eq. (14.155) are integrated forward to the guessed, transfer time t_f by using the optimal control $\hat{u} = [\lambda_z^T M(z, F)]^T / |\lambda_z^T M(z, f)|$. An iterative scheme next is used in order to adjust the initial values of the six multipliers as well as t_f such that the five terminal state parameters a_f, h_f, k_f, p_f, and q_f are matched and $(\lambda_\lambda)_f = 0$, $H_f = 1$ are satisfied. This is done by minimizing the following objective function:

$$F' = w_1(a - a_f)^2 + w_2(h - h_f)^2 + w_3(k - k_f)^2 + w_4(p - p_f)^2$$
$$+ w_5(q - q_f)^2 + w_6(\lambda_\lambda - \lambda_{\lambda_f})^2 + w_7(H - H_f)^2$$

or

$$F' = \sum_{i=1}^{5} w_i(z_i - z_{i_f})^2 + w_6(\lambda_\lambda - 0)^2 + w_7(H - 1)^2$$

with w_i standing for certain weights that can be adjusted in order to favor the rapid convergence of some elements relative to others and alleviate, to some extent, certain sensitivity and scaling problems associated with the use of a given optimizer. For this purpose, we make use of the minimization algorithm UNCMIN of Ref. 14, which is designed for the unconstrained minimization of a real-valued function $F'(x)$ of n variables denoted by the vector x. This subroutine is based on a general descent method and uses a quasi-Newton algorithm. In the Newton method, the step p'' is computed from the solution of a set of n linear equations known as the Newton equations:

$$\nabla^2 F'(x)p'' = -\nabla F'(x) \qquad (14.164)$$

Therefore, the solution is updated by using

$$x_{k+1} = x_k + p'' = x_k - \left[\nabla^2 F'(x_k)\right]^{-1} \nabla F'(x_k) \qquad (14.165)$$

Here $\nabla F'(x)$ denotes the gradient of F' at x, whereas $\nabla^2 F'(x)$, the constant matrix of the second partial derivatives of F' at x, represents the Hessian matrix. The Newton direction given by p'' is guaranteed to be a descent direction only if $[\nabla^2 F']^{-1}$ is positive definite, i.e., $z^T [\nabla^2 F']^{-1} z > 0$ for all $z \neq 0$ since, in that case, for small ε,

$$F'(x + \varepsilon p'') = F'(x) + \varepsilon \nabla F'^T p'' + \mathcal{O}(\varepsilon^2)$$

$$= F'(x) - \varepsilon \nabla F'^T [\nabla^2 F']^{-1} \nabla F' + \mathcal{O}(\varepsilon^2) \quad (14.166)$$

such that $F'(x + \varepsilon p'') < F'(x)$. This is equivalent to requiring the linear term in Newton's quadratic approximation for the function F' to be negative. This is, of course, the second term in the Taylor series expansion for F' at $x + p''$. The algorithm UNCMIN builds a secant approximation B_k to the Hessian as the function is being minimized such that, at x_k,

$$B_k p'' = -\nabla F'_k \quad (14.167)$$

with the matrix B_k positive definite since there is no guarantee that $[\nabla^2 F']^{-1}$ will always be positive definite for each x. Once the descent direction is established, a line search on α' is established such that $F'(x_k + \alpha' p'') < F'_k$ and the solution updated via $x_{k+1} = x_k + \alpha' p''$. The approximate Hessian B_k is updated next using x_{k+1} and the gradient $\nabla F'(x_{k+1})$. For example, if F' is quadratic,[14]

$$B_{k+1}(x_{k+1} - x_k) = \nabla F'_{k+1} - \nabla F'_k \quad (14.168)$$

Only the gradient values are needed for the update of the approximate Hessian, which is achieved by finite differencing such that the user must provide only the function F' itself. Let the initial orbit be given by

$$a_0 = 7000 \text{ km}, \qquad e_0 = 0, \qquad i_0 = 28.5 \text{ deg}$$

$$\Omega_0 = 0 \text{ deg}, \qquad \omega_0 = 0 \text{ deg}, \qquad M_0 = -220 \text{ deg}$$

and let $f = 9.8 \times 10^{-5}$ km/s^2 or, roughly, $10^{-2} g$. The final orbit is given by

$$a_f = 42{,}000 \text{ km}, \qquad e_f = 10^{-3}, \qquad i_f = 1 \text{ deg}$$

$$\Omega_f = 0 \text{ deg}, \qquad \omega_f = 0 \text{ deg}$$

with M_f free. The following solution is obtained:

$$(\lambda_a)_0 = 1.260484756 \text{ s/km}$$

$$(\lambda_h)_0 = 3.865626962 \times 10^2 \text{ s}$$

$$(\lambda_k)_0 = -9.388262635 \times 10^3 \text{ s}$$

$$(\lambda_p)_0 = -2.277132367 \times 10^3 \text{ s}$$

$$(\lambda_q)_0 = -1.743027218 \times 10^4 \text{ s}$$

$$(\lambda_\lambda)_0 = 5.155487187 \times 10^2 \text{ s/rad}$$

$$t_f = 58{,}624.094 \text{ s}$$

The achieved orbit parameters at $t = t_f$ are:

$$a_f = 42{,}000.007 \text{ km}, \qquad e_f = 1.00022 \times 10^{-3}$$

$$i_f = 1.000012 \text{ deg}, \qquad \Omega_f = 359.999963 \text{ deg}$$

$$\omega_f = 1.966524 \times 10^{-2} \text{ deg}, \qquad M_f = 43.779715 \text{ deg}$$

with

$$H_f = 1.002694, \qquad (\lambda_\lambda)_f = -7.662506 \times 10^{-3} \text{ s/rad}$$

The value of $\lambda_f = 19.613998$ rad is such that, with $\lambda_0 = -3.839724$ rad, a total travel of $\Delta\lambda/(2\pi) = (\lambda_f - \lambda_0)/(2\pi) = 23.453722/(2\pi) = 3.732$ revolutions around the Earth is accomplished. The initial values of the equinoctial elements are given by

$$a_0 = 7000 \text{ km}, \qquad h_0 = 0, \qquad k_0 = 0, \qquad p_0 = 0$$

$$q_0 = 0.2539676465, \qquad \lambda_0 = -3.839724354 \text{ rad}$$

Figure 14.8 shows the variations of the classical elements a and e as functions of time during the 16.284470-h transfer. The eccentricity reaches a peak of around 0.4, with most of the orbit rotation taking place in the final 2-h of the transfer. Figure 14.9 depicts e and i as functions of the semimajor axis, which is monotonically increasing as shown in Fig. 14.8. The final position given by λ_f or M_f with $M_f = 43.779715$ deg is the optimal arrival point on the target orbit that results

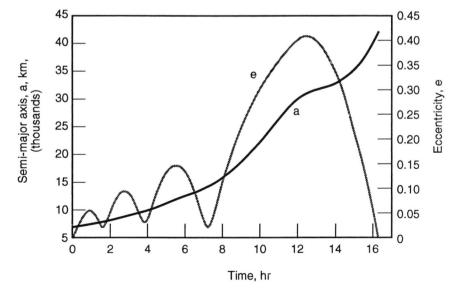

Fig. 14.8   Variation of semimajor axis and eccentricity for a LEO to a near-GEO transfer with initial $M_0 = -220$ deg (from Ref. 20).

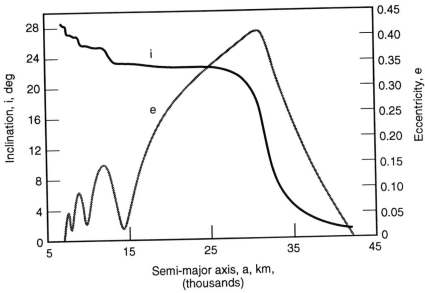

Fig. 14.9   Eccentricity and inclination vs semimajor axis for a LEO to a near-GEO transfer with initial $M_0 = -220$ deg (from Ref. 20).

in the minimum-time transfer for the given or fixed initial $M_0 = -220$ deg. The total ΔV is given by

$$\Delta V = f \cdot t_f = 5.745161 \text{ km/s}$$

We now solve the same transfer problem except that $M_0 = -300$ deg instead of -220 deg on the initial orbit at LEO. This will show whether any sensitivity to transfer time exists related to the departure location on the initial orbit. The solution now is given by

$$(\lambda_a)_0 = 7.131919334 \text{ s/km}$$

$$(\lambda_h)_0 = -6.586531878 \times 10^2 \text{ s}$$

$$(\lambda_k)_0 = 9.135791613 \times 10^3 \text{ s}$$

$$(\lambda_p)_0 = 1.439828925 \times 10^3 \text{ s}$$

$$(\lambda_q)_0 = -2.414079986 \times 10^4 \text{ s}$$

$$(\lambda_\lambda)_0 = -3.736240460 \times 10^2 \text{ s/rad}$$

$$t_f = 58{,}158.832 \text{ s}$$

This transfer requires a flight time of some 465.262 s less than the previous transfer, or some 7.75 min. The achieved parameters are

$$a_f = 42{,}000.012 \text{ km}, \qquad e_f = 9.9965 \times 10^{-4}$$

$$i_f = 1.000030 \text{ deg}, \qquad \Omega_f = 359.998924 \text{ deg}$$

$$\omega_f = 359.975378 \text{ deg}, \qquad M_f = 227.155823 \text{ deg}$$

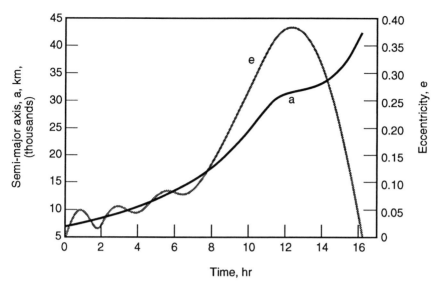

Fig. 14.10   Time history of semimajor axis and eccentricity for a LEO to a near-GEO transfer with initial $M_0 = -300$ deg (from Ref. 20).

with

$$H_f = 1.000000000$$

$$(\lambda_\lambda)_f = -8.872380 \times 10^{-3} \text{ s/rad}$$

and

$$\lambda_0 = -5.235987 \text{ rad}$$

$$\lambda_f = 16.530539 \text{ rad}$$

such that the total angular travel consists of $(\lambda_f - \lambda_0)/(2\pi) = 21.766526/(2\pi) = 3.464$ revolutions around the Earth. Figure 14.10 shows the variations of the classical orbit elements a and e during the optimal transfer with smaller oscillation amplitudes in the eccentricity during the first revolutions. Finally, Fig. 14.11 shows e and i as functions of the semimajor axis with essentially linear buildup and decay for e. The total ΔV is given by

$$\Delta V = f \cdot t_f = 5.699565 \text{ km/s}.$$

Minimum-Time Transfer with Optimized Departure and Arrival Locations

In this problem, we are given $(a)_0$, $(e)_0$, $(i)_0$, $(\Omega)_0$, and $(\omega)_0$ or, equivalently, $(a)_0$, $(h)_0$, $(k)_0$, $(p)_0$, and $(q)_0$. The guessed parameters are $(\lambda_a)_0$, $(\lambda_h)_0$, $(\lambda_k)_0$, $(\lambda_p)_0$, $(\lambda_q)_0$, and the initial mean longitude $(\lambda)_0$ as well as the transfer time t_f. The precision integration of the dynamic and adjoint systems in Eqs. (14.148–14.153)

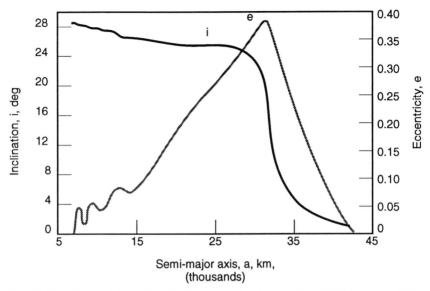

Fig. 14.11   Eccentricity and inclination vs semimajor axis for a LEO to a near-GEO transfer with initial $M_0 = -300$ deg (from Ref. 20).

and Eq. (14.155) is carried out using $(\lambda_\lambda)_0 = 0$. The boundary conditions at the unknown final time are given by $a_f, h_f, k_f, p_f, q_f, (\lambda_\lambda)_f = 0$, and $H_f = 1$ for a minimum-time solution. The optimal thrust direction still is given by

$$\hat{u} = \left[\lambda_z^T M(z, F)\right]^T / \left|\lambda_z^T M(z, F)\right|$$

The objective function to minimize still is given by F' above. We have essentially traded $(\lambda)_0$ given, with $(\lambda_\lambda)_0 = 0$, since small variations in $(\lambda)_0$ must be such that the performance index $J = -t_f$ is stationary, indicating zero sensitivity to initial location. The same argument still holds true for the arrival point on the target orbit. The solution of the optimal free-free transfer is given by

$$(\lambda_a)_0 = 4.8548563957 \text{ s/km}$$

$$(\lambda_h)_0 = 5.52370740318 \times 10^2 \text{ s}$$

$$(\lambda_k)_0 = -9.51431194293 \times 10^3 \text{ s}$$

$$(\lambda_p)_0 = -1.0373235843 \times 10^2 \text{ s}$$

$$(\lambda_q)_0 = -2.33561012603 \times 10^4 \text{ s}$$

$$(\lambda)_0 = -2.272581909 \text{ rad}$$

corresponding to $M_0 = -130.209352$ deg, and the overall minimum-time $t_f = 58,090.031$ s, with corresponding $\Delta V = 5.692823$ km/s. The total angular travel is obtained from $(\lambda)_0$ and $(\lambda)_f = 19.65399449$ rad or 3.4897 revolutions around the Earth. The final achieved parameters are:

$$a_f = 42,000.001 \text{ km}$$

$$e_f = 9.78045 \times 10^{-4}$$

$$i_f = 0.999359 \text{ deg}$$

$$\Omega_f = 358.777222 \text{ deg}$$

$$\omega_f = 350.922884 \text{ deg}$$

$$M_f = 56.390827 \text{ deg}$$

with

$$H_f = 1.038422077$$

and

$$(\lambda_\lambda)_f = -8.6830309 \times 10^{-1} \text{ s/rad}$$

The Hamiltonian is constant because the dynamic equations are not explicit functions of time. Furthermore, we can multiply all the initial values of the multipliers by a common constant factor in order to get $H = 1$ exactly, if we so desire. The solution is, of course, unchanged because of the common scaling used. This also means that we can arbitrarily select one of the λ and reduce the order of the integration by 1 since H, a positive constant, is in effect a first integral of the motion. The initial values of the multipliers that scale the Hamiltonian to the unit value are:

$$(\lambda_a)_0 = 4.675224557 \text{ s/km}$$

$$(\lambda_h)_0 = 5.319327780 \times 10^2 \text{ s}$$

$$(\lambda_k)_0 = -9.162278183 \times 10^3 \text{ s}$$

$$(\lambda_p)_0 = -9.989421518 \times 10^1 \text{ s}$$

$$(\lambda_q)_0 = -2.249191516 \times 10^4 \text{ s}$$

Figure 14.12 shows the variations of e and i as functions of the semimajor axis. Most of the inclination change is taking place at or near maximum eccentricity. The thrust pitch and yaw programs are shown in Fig. 14.13. The yaw profile, as expected, changes sign every one-half revolution during the first three orbits, rotating the orbit slowly, but the pitch angle stays near zero for maximum energy buildup. The sharp buildup in the yaw angle is responsible for most of the orbit rotation that takes place as soon as eccentricity reaches its maximum value as we have observed. Figure 14.14 shows r vs θ, the angular position with $\theta = \omega + \theta^*$. As soon as the orbit reaches the proper energy level, it is able to transfer directly to the GEO altitude on a highly eccentric orbit. The last portion of the trajectory

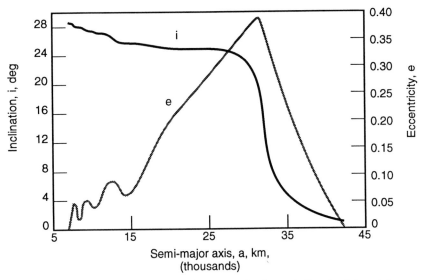

Fig. 14.12   Eccentricity and inclination vs semimajor axis for absolute minimum-time LEO to near-GEO transfer (from Ref. 20).

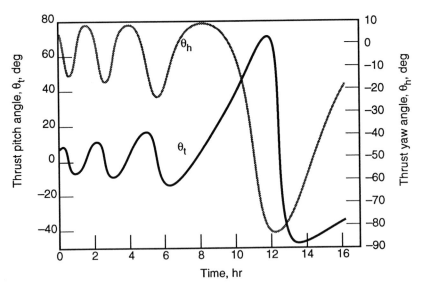

Fig. 14.13   Optimal thrust pitch and yaw programs for absolute minimum-time LEO to near-GEO transfer (from Ref. 20).

exact ΔV will be very close to the value of 5.635506 km/s since the transfer orbit will remain nearly circular throughout the exact transfer; this is the case with the averaged solution, too, which also uses a variable thrust yaw and pitch profile within each revolution during the ascent. Gravity losses are inherently more important when the orbit departs further from the circular shape as is the case with the higher acceleration of $f = 9.8 \times 10^{-5}$ km/s^2. The averaged solution fails to account effectively for the gravity losses since it provides an optimistic ΔV with respect to the exact solution, whereas the Edelbaum's ΔV_{tot} equation is overall more pessimistic, reflecting the losses associated with the use of a nonoptimal thrust profile with fixed yaw angle within each revolution. This more than compensates its inherent optimistic characteristics since it assumes a perfectly circular orbit with no gravity losses.

14.5   Orbit Transfer with Variable Specific Impulse

The problem of minimum-fuel time-fixed orbit transfer and rendezvous using continuous low thrust bounded from above and below is analyzed next. Specific impulse, or I_{sp}, is no longer considered to be a fixed quantity during the transfer and is now allowed to vary between well-defined minimum and maximum bounds such that both thrust magnitude and direction are optimized to yield the overall minimum-fuel solution. For example, for the case of the electric engine, we can assume that the power remains constant such that the thrust magnitude is inversely proportional to the specific impulse, which is continuously adjusted by varying the beam voltage. We first revisit the fundamentals of flight mechanics and low-thrust propulsion developed in Refs. 12, 17, and 18 and derive the equivalent expressions for the optimal controls for the thrust-magnitude unconstrained case using equinoctial elements instead of the usual Cartesian coordinates. The necessary conditions for optimality for the thrust-bounded case are derived following Refs. 11–13 and the problem of low-thrust transfer and rendezvous of Refs. 6, 15, and 19 extended to the case of continuously varying I_{sp}.

The Optimization of the Thrust Magnitude

From rocket propulsion fundamentals and using Newton's law for a variable mass body, the equation of motion of a rocket-powered vehicle is given by

$$m\ddot{\boldsymbol{r}} = \dot{m}\boldsymbol{c} + m\boldsymbol{g}$$

where $\boldsymbol{g}$ is the acceleration of gravity, $\boldsymbol{c}$ is the exhaust velocity, and $\dot{m} < 0$ is the rate at which mass is expelled from the engine. In this chapter, we use c for the exhaust velocity since $\hat{u}$ is used to define the thrust direction. The thrust vector $\boldsymbol{T} = \dot{m}\boldsymbol{c}$ is directed opposite the exhaust velocity vector such that the acceleration can be written as

$$f = \frac{T}{m} = \ddot{\boldsymbol{r}} - \boldsymbol{g}$$

For a low-thrust electric rocket, the exhaust stream or beam power can be written as

$$P_B = \frac{Tc}{2} \tag{14.185}$$

and, since $T = -\dot{m}c$, it can be written as

$$P_B = -\frac{T^2}{2\dot{m}} \qquad (14.186)$$

It can also be expressed in terms of the beam voltage V_B and the beam current I_B as

$$P_B = V_B I_B$$

Equation (14.185) can be derived from the following two expressions based on electrostatics considerations:

$$V_B e' = \frac{m_i}{2} c^2 \qquad (14.187)$$

$$I_B = \left(\frac{e'}{m_i}\right)(-\dot{m})$$

The first expression states that if a particle of mass m_i and charge e' and negligible initial velocity passes through a potential difference V_B, it will acquire a kinetic energy of $1/2 m_i c^2$. The second expression is the definition of the current such that the beam power P_B is written as

$$P_B = I_B V_B = \frac{1}{2}(-\dot{m})c^2$$

and, since $T = -\dot{m}c$, then,

$$P_B = \frac{1}{2}Tc$$

Alternately, from Eq. (14.187), an expression relating the I_{sp} to the beam voltage can be obtained since, with $c = I_{sp}g$,

$$c = \left[2\left(\frac{e'}{m_i}\right)V_B\right]^{1/2}$$

$$I_{sp} = \frac{1}{g}\left[2\left(\frac{e'}{m_i}\right)V_B\right]^{1/2}$$

For a given beam power P_B, the thrust vs mass flow rate curve of an electric rocket is parabolic since, from Eq. (14.186),

$$T = \sqrt{-2\dot{m}P_B} \qquad (14.188)$$

This behavior is very different from that of a constant exhaust velocity rocket since, in the latter case, the curve is linear,

$$T = -\dot{m}c$$

Conversely, the mass flow rate expressions for both types of vehicles are

$$\dot{m} = -\frac{T}{c}$$

$$\dot{m} = \frac{-T^2}{2P_B} \tag{14.189}$$

It is therefore advantageous from a propellant consumption point of view to have high exhaust velocity or high power. Furthermore, Eq. (14.188) shows that the same level of thrust can be achieved by different combinations of $\dot{m}$ and P_B since, if a lower power level is selected, an appropriate increase in $\dot{m}$ will maintain T constant. However, from Eq. (14.189) and for a given T, it is seen that $\dot{m}$ is at a minimum if P_B is chosen at its maximum level. This means that, at each instant of time, the selection of P_B at its maximum value, namely, $P_{B_{\max}}$, will achieve the required thrust for minimum $\dot{m}$ or minimum fuel consumption. In other words, from all the possible ways of flying a required trajectory, meaning an acceleration time history $f(t) = \ddot{r}(t) - g(t)$, the selection of $P_{B_{\max}}$ is the only one that results in minimum fuel expenditure.[12] Electric rockets must therefore always operate at $P_{B_{\max}}$. Then, from Eq. (14.185) with $P_B = P_{B_{\max}}$, as the thrust is decreased, $I_{\rm sp}$ will increase and vice versa. Since the $I_{\rm sp}$ is dependent on the beam voltage V_B, it can be obtained by adjusting V_B, provided that I_B is also adjusted to obtain the required $P_{B_{\max}}$ power since $P_B = I_B V_B$. All these fundamental ideas are encapsulated in Figs. 14.18–14.20. In Fig. 14.18, the thrust vs mass flow rate curve for a low-thrust rocket with constant exhaust velocity is depicted. Since c cannot be varied in this type of rocket, the thruster operates at $T_{\max}$ only.[12]

Figure 14.19 corresponds to the variable c case with $T = (-2\dot{m}P)^{1/2}$ for given power P_B or for short, P. The power levels below $P_{\max}$ result in Region II, with Region I completely inaccessible since it corresponds to $P > P_{\max}$, which is impossible. As is pointed out by Marec in Ref. 12, it is not optimal to operate

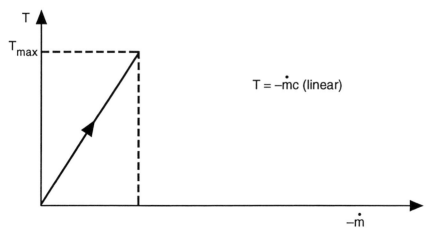

Fig. 14.18   Thrust vs mass flow rate for constant–exhaust velocity low-thrust rocket (from Ref. 12).

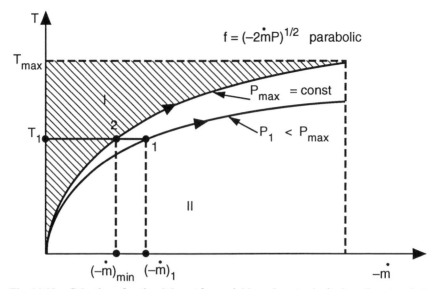

Fig. 14.19   Selection of optimal thrust for variable–exhaust velocity low-thrust rocket (from Ref. 12).

these types of thrusters at a power level $P_1 < P_{max}$ since, for the same thrust T_1, the operation at P_{max} depicted by point 2 results in the minimum mass flow rate or propellant expenditure. In Fig. 14.20, and for given P_{max}, any operating point A corresponds to a unique combination of beam voltage V_B and beam current I_B. If V_B is increased, then I_{sp} or c will also increase according to the I_{sp} equation and, since power is held at $P_{max} = $ constant, the thrust will decrease accordingly since $P = 1/2Tc$. Conversely, decreasing I_{sp} or V_B will result in increasing thrust and I_B. In Fig. 14.21, the unreachable Region I is extended further by the inclusion of the boundary OO', which corresponds to the equation $T = -\dot{m}c_{max}$, where c_{max} is the maximum exhaust velocity achieved by the rocket. This boundary is necessary to prevent the exhaust velocity or the I_{sp} to grow to very large values as the thrust is decreased toward its minimum value. This minimum is conveniently defined at point O such that the operating arc is the arc O3 on the P_{max} parabola. Following Edelbaum and letting r and v stand for the spacecraft position and velocity vectors, the second-order differential equation of motion is reduced to the following first-order form:

$$\dot{v} = \frac{T}{m} + g(r, t)$$

$$\dot{r} = v$$

The mass flow rate obeys the general form

$$\dot{m} = \dot{m}(r, t, T)$$

The optimal value of the specific impulse is then given by

$$I_{sp}^* = \frac{2P}{T^* g}$$

since $c = 2P/T = I_{sp}g$. Now the optimal acceleration program is obtained from

$$f^* = \frac{T^*}{m} = \frac{\lambda_z^T M \hat{u}}{m^2 \lambda_m} P$$

If the thrust magnitude is bounded from above and below, then,

$$T_{min} < T < T_{max}$$

such that we have the following inequality constraints on the control variable T:

$$C_1 = T - T_{max} \leq 0$$
$$C_2 = -T + T_{min} \leq 0$$

These constraints can be adjoined to the original Hamiltonian by way of Lagrange multipliers μ_1 and μ_2 such that

$$H = \frac{T}{m}\lambda_z^T M \hat{u} - \frac{T^2}{2P}\lambda_m + \lambda_\lambda n + \mu_1(T - T_{max}) + \mu_2(-T + T_{min})$$

The necessary condition on H is

$$H_T = \frac{\partial H}{\partial T} = \lambda_z^T \frac{M}{m}\hat{u} - \frac{T}{P}\lambda_m + \mu_1 - \mu_2 = 0$$

The multipliers μ_1 and μ_2 are such that $\mu_1 > 0$ when $C_1 = 0$ or $T = T_{max}$, and $\mu_1 = 0$ when $C_1 < 0$ or $T < T_{max}$ and, similarly, $\mu_2 > 0$ when $C_2 = 0$ or $T = T_{min}$, and $\mu_2 = 0$ when $C_2 < 0$ or $T > T_{min}$. When $T_{min} < T < T_{max}$ assumes an intermediate value, $\mu_1 = \mu_2 = 0$ and $H_T = 0$ reduces to $\lambda_z^T (M/m)\hat{u}$ $-(T/P)\lambda_m = 0$, yielding the optimal control given by Eq. (14.201). The values of μ_1 and μ_2 are obtained from

$$\mu_1 = \frac{T_{max}}{P}\lambda_m - \frac{\lambda_z^T M \hat{u}}{m}$$

$$\mu_2 = -\frac{T_{min}}{P}\lambda_m + \frac{\lambda_z^T M \hat{u}}{m}$$

The Lagrange multipliers are still given by Eqs. (14.198) and (14.199), and the optimal T^* is selected by monitoring the value of $\lambda_z^T M \hat{u} P/(m\lambda_m)$. If it is less than T_{min}, then $T^* = T_{min}$ and, if it is larger than T_{max}, then $T^* = T_{max}$ and, finally, if it is intermediate between T_{min} and T_{max}, then T^* is given by Eq. (14.201), which is the presently calculated value of $\lambda_z^T M \hat{u} P/(m\lambda_m)$. We can also use the simpler Hamiltonian H^* without adjoining the constraints, namely,

$$H^* = \lambda_z^T \dot{z} + \lambda_m \dot{m} = \frac{T}{m}\left(\lambda_z^T M \hat{u} - \frac{m\lambda_m}{c}\right) + \lambda_\lambda n$$

This is equivalent to Eq. (14.202) since $c = 2P/T$ is a function of the control T,

$$H^* = \frac{T}{m}\lambda_z^T M \hat{u} - \frac{T^2}{2P}\lambda_m + \lambda_\lambda n \qquad (14.202)$$

The optimality condition yields with

$$\frac{\partial H^*}{\partial T} = H_T^* = \lambda_z^T \frac{M}{m}\hat{u} - \frac{T}{P}\lambda_m \qquad (14.203)$$

the following condition: $\delta H^* = H_T^* \delta T \leq 0$ since this is equivalent to $\delta J = \int_{t_0}^{t_f} H_T^* \delta T \, dt = \int_{t_0}^{t_f} \delta H^* dt \leq 0$ for the control T to be maximizing for all admissible values of δT. δJ is the variation in J, the performance index, due to variations in T for fixed $z(t_0)$. The optimal control is selected by monitoring the value of H_T^*. If H_T^*, as calculated by Eq. (14.203), is positive, then $T = T_{\max}$ and, if H_T^* is negative, then $T = T_{\min}$ and, finally, if $H_T^* = 0$ then T^* is given by Eq. (14.203):

$$H_T^* > 0 \Rightarrow \delta T < 0 \Rightarrow T = T_{\max} \qquad (14.204)$$

$$H_T^* < 0 \Rightarrow \delta T > 0 \Rightarrow T = T_{\min} \qquad (14.205)$$

$$H_T^* = 0 \Rightarrow T = T^* = \lambda_z^T \frac{M\hat{u}}{m\lambda_m}P \qquad (14.206)$$

In practice, the last condition for $H_T^* = 0$ is replaced by $\left|H_T^*\right| < \varepsilon$, where ε is a small number, say 10^{-10}. The Euler–Lagrange equations are still given by Eqs. (14.198) and (14.199). We now maximize the value of the mass at the fixed final time t_f such that the performance index $J = \phi = m_f$ and the optimal thrust direction given by Eq. (14.200) is obtained directly from the maximum principle. Since the equations of motion given by Eqs. (14.192) and (14.193) are not explicit functions of time, the Hamiltonian H in Eq. (14.194), or H^* in Eq. (14.202), is constant throughout the transfer. Given initial state parameters $(a)_0, (h)_0, (k)_0, (p)_0, (q)_0, (\lambda)_0$, and $(m)_0$, the initial values of the seven Lagrange multipliers are guessed, namely, $(\lambda_a)_0, (\lambda_h)_0, (\lambda_k)_0, (\lambda_p)_0, (\lambda_q)_0, (\lambda_\lambda)_0, (\lambda_m)_0$, and the state and adjoint equations given by Eqs. (14.192), (14.193) and (14.198), (14.199) are integrated forward from t_0 to t_f by using the optimal thrust direction $\hat{u}$ in Eq. (14.200) and the thrust magnitude from Eqs. (14.204–14.206). The initial values of the multipliers are iterated until the desired terminal state given by $(a)_f, (h)_f, (k)_f, (p)_f, (q)_f, (\lambda)_f$, and $(\lambda_m)_f = (\partial\phi/\partial m)_{t_f} = 1$ is satisfied. This is achieved by minimizing the following objective function:

$$F' = w_1[a - (a)_f]^2 + w_2[h - (h)_f]^2 + w_3[k - (k)_f]^2$$

$$+ w_4[p - (p)_f]^2 + w_5[q - (q)_f]^2 + w_6[\lambda - (\lambda)_f]^2 + w_7[\lambda_m - 1]^2$$

$$\frac{\partial M_{43}}{\partial a} = \frac{(1 + p^2 + q^2)}{2na^2(1 - h^2 - k^2)^{1/2}} \left(-\frac{1}{2a} Y_1 + \frac{\partial Y_1}{\partial a} \right) \tag{A.66}$$

$$\frac{\partial M_{51}}{\partial a} = 0 \tag{A.67}$$

$$\frac{\partial M_{52}}{\partial a} = 0 \tag{A.68}$$

$$\frac{\partial M_{53}}{\partial a} = \frac{(1 + p^2 + q^2)}{2na^2(1 - h^2 - k^2)^{1/2}} \left(-\frac{1}{2a} X_1 + \frac{\partial X_1}{\partial a} \right) \tag{A.69}$$

$$\frac{\partial M_{61}}{\partial a} = \frac{-M_{61}}{2a} + \frac{1}{na^2}$$
$$\times \left[-2\frac{\partial X_1}{\partial a} + (1 - h^2 - k^2)^{1/2} \left(h\beta \frac{\partial^2 X_1}{\partial a \partial h} + k\beta \frac{\partial^2 X_1}{\partial a \partial k} \right) \right] \tag{A.70}$$

$$\frac{\partial M_{62}}{\partial a} = -\frac{M_{62}}{2a} + \frac{1}{na^2}$$
$$\times \left[-2\frac{\partial Y_1}{\partial a} + (1 - h^2 - k^2)^{1/2} \left(h\beta \frac{\partial^2 Y_1}{\partial a \partial h} + k\beta \frac{\partial^2 Y_1}{\partial a \partial k} \right) \right] \tag{A.71}$$

$$\frac{\partial M_{63}}{\partial a} = -\frac{M_{63}}{2a} + \frac{1}{na^2} \left[\left(q \frac{\partial Y_1}{\partial a} - p \frac{\partial X_1}{\partial a} \right) (1 - h^2 - k^2)^{-1/2} \right] \tag{A.72}$$

with

$$\frac{\partial \dot{X}_1}{\partial a} = -\frac{1}{2} \frac{na}{r} \left[hk\beta c_F - (1 - h^2\beta)s_F \right] \tag{A.73}$$

$$\frac{\partial \dot{Y}_1}{\partial a} = \frac{1}{2} \frac{na}{r} \left[hk\beta s_F - (1 - k^2\beta)c_F \right] \tag{A.74}$$

The Partial Derivatives of M with Respect to λ

$$\frac{\partial M_{11}}{\partial \lambda} = \frac{2}{n^2 r} \frac{\partial \dot{X}_1}{\partial F} \tag{A.75}$$

$$\frac{\partial M_{12}}{\partial \lambda} = \frac{2}{n^2 r} \frac{\partial \dot{Y}_1}{\partial F} \tag{A.76}$$

$$\frac{\partial M_{13}}{\partial \lambda} = 0 \tag{A.77}$$

$$\frac{\partial M_{21}}{\partial \lambda} = \frac{(1 - h^2 - k^2)^{1/2}}{nar} \left(\frac{\partial^2 X_1}{\partial F \partial k} - \frac{h\beta}{n} \frac{\partial \dot{X}_1}{\partial F} \right) \tag{A.78}$$

$$\frac{\partial M_{22}}{\partial \lambda} = \frac{(1 - h^2 - k^2)^{1/2}}{nar} \left(\frac{\partial^2 Y_1}{\partial F \partial k} - \frac{h\beta}{n} \frac{\partial \dot{Y}_1}{\partial F} \right) \tag{A.79}$$

$$\frac{\partial M_{23}}{\partial \lambda} = \frac{k \left(q \frac{\partial Y_1}{\partial F} - p \frac{\partial X_1}{\partial F} \right)}{nar(1 - h^2 - k^2)^{1/2}} \tag{A.80}$$

$$\frac{\partial M_{31}}{\partial \lambda} = -\frac{(1 - h^2 - k^2)^{1/2}}{nar} \left(\frac{\partial^2 X_1}{\partial F \partial h} + \frac{k\beta}{n} \frac{\partial \dot{X}_1}{\partial F} \right) \tag{A.81}$$

$$\frac{\partial M_{32}}{\partial \lambda} = -\frac{(1 - h^2 - k^2)^{1/2}}{nar} \left(\frac{\partial^2 Y_1}{\partial F \partial h} + \frac{k\beta}{n} \frac{\partial \dot{Y}_1}{\partial F} \right) \tag{A.82}$$

$$\frac{\partial M_{33}}{\partial \lambda} = \frac{-h \left(q \frac{\partial Y_1}{\partial F} - p \frac{\partial X_1}{\partial F} \right)}{nar(1 - h^2 - k^2)^{1/2}} \tag{A.83}$$

$$\frac{\partial M_{41}}{\partial \lambda} = \frac{\partial M_{42}}{\partial \lambda} = 0 \tag{A.84}$$

$$\frac{\partial M_{43}}{\partial \lambda} = \frac{(1 + p^2 + q^2)}{2nar(1 - h^2 - k^2)^{1/2}} \frac{\partial Y_1}{\partial F} \tag{A.85}$$

$$\frac{\partial M_{51}}{\partial \lambda} = \frac{\partial M_{52}}{\partial \lambda} = 0 \tag{A.86}$$

$$\frac{\partial M_{53}}{\partial \lambda} = \frac{(1 + p^2 + q^2)}{2nar(1 - h^2 - k^2)^{1/2}} \frac{\partial X_1}{\partial F} \tag{A.87}$$

$$\frac{\partial M_{61}}{\partial \lambda} = \frac{1}{nar} \left[-2\frac{\partial X_1}{\partial F} + (1 - h^2 - k^2)^{1/2} \left(h\beta \frac{\partial^2 X_1}{\partial F \partial h} + k\beta \frac{\partial^2 X_1}{\partial F \partial k} \right) \right] \tag{A.88}$$

$$\frac{\partial M_{62}}{\partial \lambda} = \frac{1}{nar} \left[-2\frac{\partial Y_1}{\partial F} + (1 - h^2 - k^2)^{1/2} \left(h\beta \frac{\partial^2 Y_1}{\partial F \partial h} + k\beta \frac{\partial^2 Y_1}{\partial F \partial k} \right) \right] \tag{A.89}$$

$$\frac{\partial M_{63}}{\partial \lambda} = \frac{\left(q \frac{\partial Y_1}{\partial F} - p \frac{\partial X_1}{\partial F} \right)}{nar(1 - h^2 - k^2)^{1/2}} \tag{A.90}$$

The auxiliary partials are

$$\frac{\partial X_1}{\partial F} = a\left[hk\beta c_F - (1 - h^2\beta)s_F \right] \tag{A.91}$$

$$\frac{\partial Y_1}{\partial F} = a\left[-hk\beta s_F + (1 - k^2\beta)c_F \right] \tag{A.92}$$

$$\frac{\partial \dot{X}_1}{\partial F} = -\frac{a}{r}(ks_F - hc_F)\dot{X}_1 + \frac{a^2 n}{r} \left[-hk\beta s_F - (1 - h^2\beta)c_F \right] \tag{A.93}$$

$$\frac{\partial \dot{Y}_1}{\partial F} = -\frac{a}{r}(ks_F - hc_F)\dot{Y}_1 + \frac{a^2 n}{r}\left[-hk\beta c_F - (1 - k^2\beta)s_F \right] \quad \text{(A.94)}$$

$$\frac{\partial^2 X_1}{\partial F \partial h} = a\left[(hs_F + kc_F)\left(\beta + \frac{h^2\beta^3}{1 - \beta} \right) + \frac{a^2}{r^2}(h\beta - s_F)(s_F - h) + \frac{a}{r}c_F^2 \right]$$
$$\text{(A.95)}$$

$$\frac{\partial^2 X_1}{\partial F \partial k} = -a\left[-(hs_F + kc_F)\frac{hk\beta^3}{1 - \beta} + \frac{a^2}{r^2}(s_F - h\beta)(c_F - h) + \frac{a}{r}s_F c_F \right]$$
$$\text{(A.96)}$$

$$\frac{\partial^2 Y_1}{\partial F \partial h} = a\left[-(hs_F + kc_F)\frac{hk\beta^3}{1 - \beta} - \frac{a^2}{r^2}(k\beta - c_F)(s_F - h) + \frac{a}{r}s_F c_F \right]$$
$$\text{(A.97)}$$

$$\frac{\partial^2 Y_1}{\partial F \partial k} = a\left[-(hs_F + kc_F)\left(\beta + \frac{k^2\beta^3}{1 - \beta} \right) + \frac{a^2}{r^2}(c_F - k\beta)(c_F - k) - \frac{a}{r}s_F^2 \right]$$
$$\text{(A.98)}$$

References

[1]Edelbaum, T. N., "Propulsion Requirements for Controllable Satellites," *ARSJ*, Aug. 1961, pp. 1079–89.

[2]Broucke, R. A., and Cefola, P. J., "On the Equinoctial Orbit Elements," *Celestial Mechanics* 5, pp. 303–310, 1972.

[3]Herrick, S., *Astrodynamics*, Vol. II, Van Nostrand Reinhold, London, 1972.

[4]Plummer, H. C., *An Introductory Treatise on Dynamical Astronomy*, Dover, New York, 1960.

[5]Cefola, P. J., "Equinoctial Orbit Elements: Application to Artificial Satellite Orbits," AIAA Paper 72-937, AIAA/AAS Astrodynamics Conference, Palo Alto, CA, Sept. 11–12, 1972.

[6]Edelbaum, T. N., Sackett, L. L., and Malchow, H. L., "Optimal Low Thrust Geocentric Transfer," AIAA Paper 73-1074, AIAA 10th Electric Propulsion Conference, Lake Tahoe, NV, Oct. 31–Nov. 2, 1973.

[7]Cefola, P. J., Long, A. C., and Holloway, G., Jr., "The Long-Term Prediction of Artificial Satellite Orbits," AIAA Paper 74-170, AIAA 12th Aerospace Sciences Meeting, Washington, DC, Jan. 30–Feb. 1, 1974.

[8]Sackett, L. L., and Edelbaum, T. N., "Effect of Attitude Constraints on Solar-Electric Geocentric Transfers," AIAA Paper 75-350, AIAA 11th Electric Propulsion Conference, New Orleans, LA, March 19–21, 1975.

[9]Kechichian, J. A., "Equinoctial Orbit Elements: Application to Optimal Transfer Problems," AIAA Paper 90-2976, AIAA/AAS Astrodynamics Conference, Portland, OR, Aug. 20–22, 1990.

[10]Kechichian, J. A., "Trajectory Optimization with a Modified Set of Equinoctial Orbit Elements," AAS/AIAA Paper 91-524, Astrodynamics Specialist Conference, Durango, CO, Aug. 19–22, 1991.

[11]Bryson, A. E., and Ho, Y-C., *Applied Optimal Control,* Ginn, Waltham, MA, 1969.

[12]Marec, J-P., *Optimal Space Trajectories,* Elsevier, Amsterdam, 1979.

[13]Vinh, N. X., *Optimal Trajectories in Atmospheric Flight,* Elsevier, Amsterdam, 1981.

[14]Kahaner, D., Moler, C., and Nash, S., *Numerical Methods and Software,* Prentice-Hall, Englewood Cliffs, NJ, 1989.

[15]Kechichian, J. A., "Optimal Low-Thrust Rendezvous Using Equinoctial Orbit Elements," IAF Paper 92-0014, 43rd Congress of the International Astronautical Federation, Washington, DC, Aug. 28–Sept. 5, 1992.

[16]Kechichian, J. A., "Optimal Low-Thrust Transfer Using Variable Bounded Thrust," IAF Paper 93A.2.10, 44th Congress of the International Astronautical Federation, Graz, Austria, Oct. 16–22, 1993.

[17]Edelbaum, T. N., "Optimal Space Trajectories," Analytic Mechanics Associates, Inc., Jericho, NY, Dec. 1969.

[18]Hill, P. G., and Peterson, C. R., *Mechanics and Thermodynamics of Propulsion,* Addison-Wesley, Reading, MA, 1970.

[19]Kechichian, J. A., "Minimum-Fuel Time-Fixed Rendezvous Using Constant Low Thrust," AAS Paper 93-130, AAS/AIAA Spaceflight Mechanics Meeting, Pasadena, CA, Feb. 22–24, 1993.

[20]Kechichian, J. A., "Optimal LEO-GEO Intermediate Acceleration Orbit Transfer," AAS Paper 94-125, AAS/AIAA Spaceflight Mechanics Meeting, Cocoa Beach, FL, Feb. 14–16, 1994.

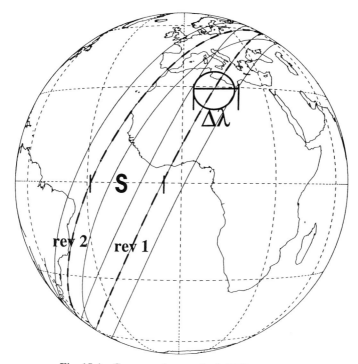

Fig. 15.4   Coverage from a single LEO satellite.

where

a_0 = initial semimajor axis
e_0 = initial eccentricity
i_0 = initial inclination
ω_0 = initial argument of perigee

are the osculating elements at the ascending node and

$$P_0 = \text{Keplerian period} = 2\pi \left(a_0^3/\mu\right)^{0.5}$$

It is useful in examining subsequent revolutions of a satellite to define the dimensionless satellite trace repetition parameter Q:

$$Q = 360/S \qquad (15.8)$$

Note that S is measured positively westward. From Fig. 15.4 it can be seen that, if the westward shift S of each revolution is such that it divides into 360 deg an integral number of times, then the $Q + 1$ pass will have the same ascending node longitude as the first pass. Therefore, if Q is an integer, the orbit is called a *repeating groundtrack orbit*, whose groundtrack repeats itself each day after Q revolutions. A geosynchronous orbit (see Sec. 11.3) is a $Q = 1$ orbit since its groundtrack repeats daily after a single revolution. A Molniya orbit (see Sec. 11.4) is an example of a $Q = 2$ orbit. Its groundtrack repeats daily after two revolutions.

In general, Q need not be an integer but can be expressed as a ratio of integers:

$$Q = \frac{N}{D} \tag{15.9}$$

In this form, D is the number of days until the groundtrack repeats itself, and N is the number of orbit revolutions until repeat. As an example, a $Q = 12\frac{1}{3} = 37/3$ orbit will repeat the same groundtrack every three days, after 37 revolutions. A $Q = 12.3 = 123/10$ orbit will repeat the same groundtrack every 10 days, after 123 revolutions. Even though the groundtrack of this latter orbit does repeat, it would probably not be considered a repeating groundtrack orbit since the repeat interval is so long.

The satellite trace repetition parameter Q can be used to analyze how successive passes and their coverage swaths overlay on the Earth's surface. For satellites at the same altitude, it can be used to interleave their passes to achieve the desired coverage.

At any given latitude (ϕ), the longitude region ($\Delta\lambda$), as shown in Fig. 15.4 swept out by the swath, can be found from

$$\Delta\lambda = \sin^{-1}\left(\frac{\sin\theta + \sin\phi\cos i}{\sin i\cos\phi}\right) + \sin^{-1}\left(\frac{\sin\theta - \sin\phi\cos i}{\sin i\cos\phi}\right) \tag{15.10}$$

for values of latitude ϕ such that $-i \mp \theta \leq \phi \leq i \pm \theta$ (where the upper sign is for posigrade inclinations and the lower sign is for retrograde inclinations). When the swath width θ is small, this equation becomes

$$\Delta\lambda \approx \frac{2\theta}{(\cos^2\phi - \cos^2 i)^{0.5}} \tag{15.11}$$

For the latitude value in which the longitude swept out is equal to the distance between groundtracks (i.e., $\Delta\lambda = S$), there will be no gaps between swaths of successive revolutions. The entire region above this latitude and up to a latitude near ($i + \theta$) for posigrade orbits will be covered by the satellite in a single day. In fact, it will be covered at least twice, once by northbound passes and once by southbound passes.

Clearly, the equator is the most difficult latitude to cover using highly inclined orbits. Complete coverage of the equator requires a coverage circle size given by

$$\theta = \sin^{-1}\left(\sin\frac{S}{2}\sin i\right) \tag{15.12}$$

Coverage of the poles is assured so long as $i + \theta \geq 90$ (or $i - \theta \leq 90$ for retrograde orbits). In this case, global coverage would be achieved. The revisit time at the equator would be about 12 h, with shorter revisit times toward the poles. A given point on Earth would see the satellite two or more times per day, but each viewing opportunity would be only minutes in duration.

Reducing the revisit time could be accomplished by using additional LEO satellites that come into view of the region of interest at other times of day so as to break up large coverage gap intervals. Hanson et al.[1] have studied methods for selecting the initial values of longitude of ascending nodes and mean anomaly to minimize revisit time for a constellation of satellites to a point on the Earth. To achieve continuous global coverage using LEO satellites obviously requires constellations of many satellites. This subject will be addressed later in the chapter.

The LEO orbit is a good choice for Earth resources, weather, or surveillance satellites, which do not require continuous or even quick revisit time coverage. In addition, sensors with limited slant range capability find the LEO orbit a necessity. The LEO orbit also has the advantage of requiring the least energy to achieve (see Fig. 5.2). Hence, it is favored for extremely heavy satellites (e.g., Space Shuttle, Space Station, Hubble Telescope) and those seeking to launch on small launch vehicles (e.g., Pegasus).

Coverage from a Single Geosynchronous Equatorial Orbit Satellite

By definition, a geosynchronous satellite revolves in its orbit at the same rate at which the Earth rotates about its polar axis. If the orbital plane of the satellite is equatorial ($i = 0$), the satellite remains over the same point on the equator. Its groundtrack is simply a point on the Earth.

Figure 15.5 shows the Earth coverage from a single geosynchronous equatorial orbit (GEO) satellite. Because the satellite does not move relative to the Earth, it has the same view continuously. Contours of different ground elevation angle (ε) are shown on the figure. Because the satellite is at high altitude ($h = 35{,}786\,\text{km}$), it can see a large region of the Earth continuously. The GEO orbit is an excellent choice for continuous coverage of nearly a hemisphere of the Earth. A single satellite can provide 24-h communication coverage for the North and South American continents (excluding the extreme polar regions). A single satellite could also continuously link most of Europe, Africa, South America, and North America as shown in Fig. 15.5. A set of three GEO satellites would continuously cover all but the polar regions as shown in Fig. 15.6. The drawbacks of the GEO orbit are the large amount of energy required to achieve it (see Fig. 5.2), which translates into high launch costs, and the large sensor range required.

A geosynchronous satellite that has a nonzero inclination will trace out a figure eight on the Earth. The maximum latitude excursion of the subsatellite point will equal the orbital inclination. Figure 15.7 shows the groundtrack and coverage for such a satellite with an inclination of 60 deg. The contours show the regions that are covered 24, 18, 12, 6, and 0 h/day. Because the satellite is moving relative to the Earth's surface, it has visibility to a greater region than the motionless equatorial GEO satellite did. The only region that this satellite never sees is the small football-shaped region 180 deg away from its groundtrack on the Earth. On the other hand, the region that is always in view, namely, the small football-shaped region centered on the groundtrack, is much smaller for the inclined GEO. So, although, the inclined GEO covers more of the Earth, it covers less of it continuously (24 h/day). Unlike the equatorial GEO, an inclined geosynchronous orbit would allow access to the polar regions for significant time periods each day. The North Pole, which could not be seen by the equatorial GEO, can be covered about 10 h/day by the inclined GEO.

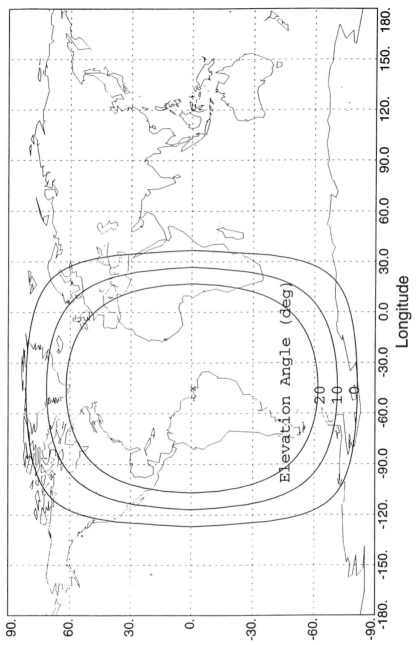

Fig. 15.5   Coverage from a single GEO satellite.

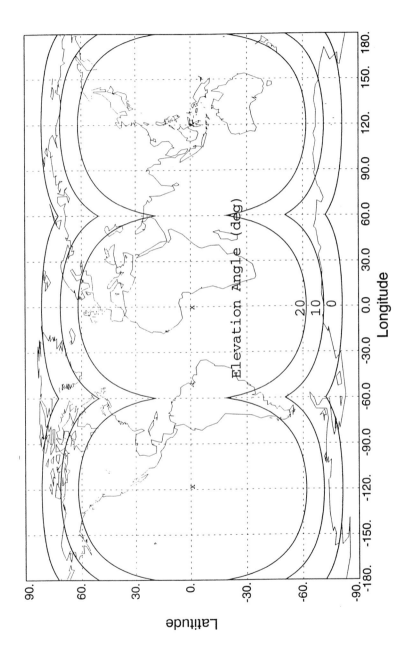

Fig. 15.6   Coverage from three GEO satellites.

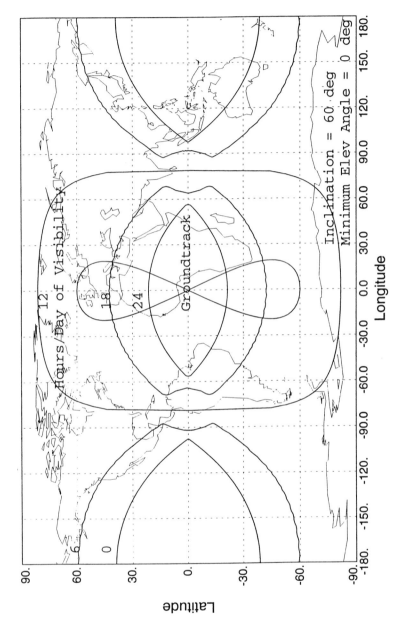

Fig. 15.7   Coverage from one inclined GEO satellite.

Coverage of the polar regions can be further enhanced by using inclined, eccentric GEO orbits. An example is shown in Fig. 15.8, for a geosynchronous satellite with an inclination of 30 deg, an eccentricity of 0.3, and an argument of perigee of 270 deg. (Note that, because the inclination is not at the critical value of 63.435 deg, the argument of perigee will increase slowly with time as a result of the effects of J_2.) This satellite traverses a nearly circular groundtrack in a counterclockwise direction, with apogee at the northernmost point and perigee at the southernmost point. As a result, the satellite spends more time north of the equator than south of it. Consequently, the coverage is shifted toward the northern hemisphere. The North Pole is now covered about 14 h/day compared to about 5 h/day for the South Pole. Again, the region covered continuously (24 h) is smaller than for the circular, equatorial GEO.

Combinations of GEO satellites with and without inclination and eccentricity can be used effectively to provide regional, global, or even polar coverage. Hanson and Higgins[2] examined such combinations of GEO satellites to maximize coverage of six different geographic areas. Their results show that, for global or near-global coverage, constellations of elliptical or circular GEO satellites perform about equally well. For coverage of the northern hemisphere or of a region such as the United States, North Atlantic, and Western Europe, the elliptical GEO constellations offer the better coverage.

Coverage from a Single Highly Eccentric Orbit Satellite

In the previous section, it was noted that eccentric orbits with apogee located at the northernmost point in the orbit could be used to shift coverage to favor the northern hemisphere. In the current section, this concept is taken nearly to extreme in the study of the highly eccentric orbit (HEO).

In a highly eccentric orbit, the satellite spends most of its time in the region of apogee at a high-altitude vantage point, where it sees the largest surface area of the Earth. Relatively little time is spent at low altitudes because the satellite speeds through perigee on its way back out toward apogee. By far the most common HEO is the highly eccentric, critically inclined $Q = 2$ (Molniya) orbit that was examined in Sec. 11.4. The typical Molniya orbit has an apogee altitude higher than GEO and a perigee altitude in the 900- to 1800-km range. Its orbital period of 11.967 h is one-half of a mean sidereal day ($Q = 2$), so that the groundtrack of the satellite repeats itself every two revolutions. The critical inclination of 63.435 deg is employed to prevent rotation of the line of apsides so as to maintain apogee at the northernmost point in the orbit.

The instantaneous view of the Earth from a Molniya satellite at apogee is shown in Fig. 15.9. The groundtrack and coverage provided by a single Molniya satellite are shown in Fig. 15.10. The groundtrack is labeled with two apogees, 180° apart in longitude, and two perigees similarly spaced. In a single day, the satellite traverses the groundtrack from west to east and returns to repeat the same groundtrack the next day. Since the satellite spends 11 h of its 12-h period north of the equator (most of that time in the vicinity of apogee), the coverage is concentrated in the northernmost regions. Notice that a significant amount of the northern region is covered 6, 9, and even 12 h/day. In cases where extended coverage of a polar region (typically northern) is desired, the Molniya orbit is an excellent choice. It

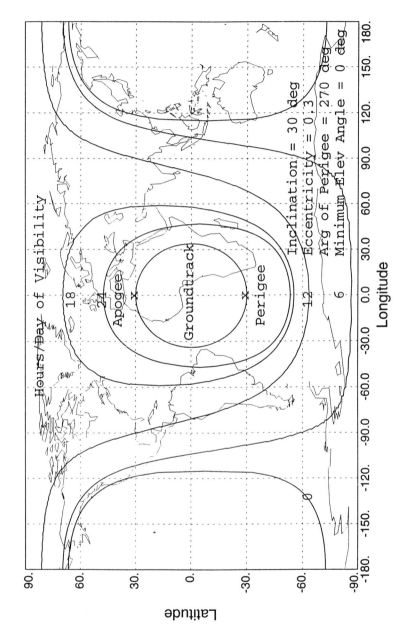

Fig. 15.8   Coverage from one inclined eccentric GEO.

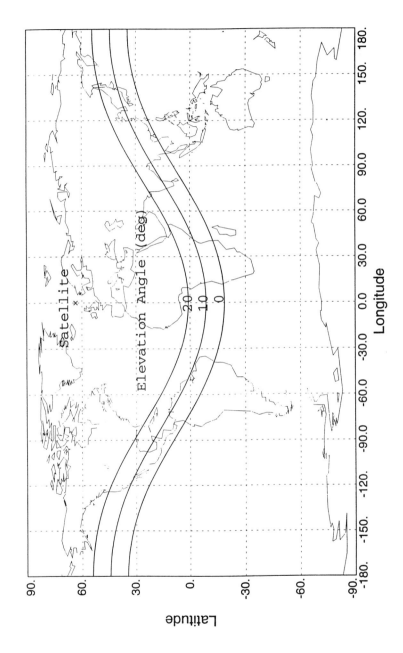

Fig. 15.9   Instantaneous view by HEO at apogee.

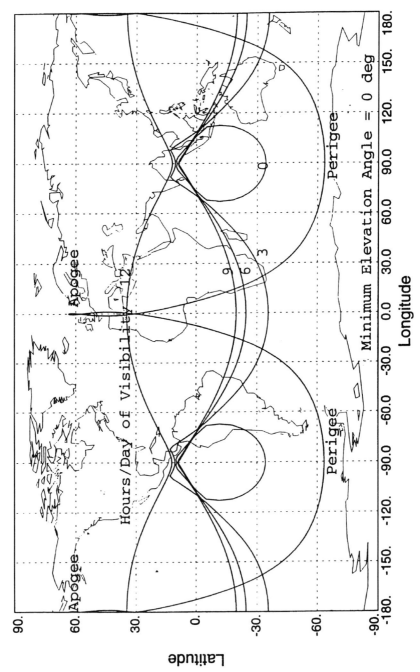

Fig. 15.10   Coverage from a single HEO satellite.

requires less energy to achieve than does the GEO orbit (see Fig 5.2), because perigee remains at low altitude.

For continuous viewing of northern regions, constellations of two or three Molniya satellites are often used. For a two-satellite arrangement, the second satellite is placed in the same groundtrack as the first but phased so that one satellite is at perigee while the other is at apogee. To do this requires using orbital planes 90 deg apart in right ascension of ascending node and phasings (mean anomalies) that differ by 180 deg. For a three-satellite arrangement, all three satellites are placed in the same groundtrack but phased 8 h (240 deg in mean anomaly) apart. The right ascensions of ascending node in this case are spaced 120 deg apart. The continuous and partial coverages from these two- and three-satellite constellations are shown in Figs. 15.11 and 15.12, respectively. The two-satellite HEO constellation can continuously cover most of the Earth above 30° north latitude. The three-satellite HEO constellation can continuously cover nearly all of the northern hemisphere. Clearly, these constellations of HEO satellites are quite efficient at concentrating coverage in the northern (or southern if apogee is placed at the southernmost point in the orbit) regions of the Earth.

Coverage from a Single Medium Earth Orbit Satellite

Although there is no strict definition of what constitutes a medium earth orbit (MEO), it is safe to say that any orbit too high to be labeled LEO, too low for GEO, and not specifically an HEO is an MEO. Typically, the MEO label is applied to orbits whose periods range from about 2 to 18 h and whose eccentricity is small. This range includes orbit altitudes in the 2000- to 30,000-km region. If the LEO, with its small coverage circle, is at one extreme of the coverage realm and the GEO, with its nearly hemispheric coverage circle, is at the other, then the MEO falls in between.

In Fig. 15.3, the LEO orbits are near the left end of the plot, and the GEO orbits are near the right end of the plot. The MEO orbits constitute the rest. In Fig. 15.3, note how quickly the size of the coverage circle increases as the orbital altitude is increased from 2000 to 10,000 km. Beyond this point, the coverage circle size increases only moderately with altitude. The MEO can offer the orbit planner a middle ground between the LEO and GEO alternatives. The MEO offers a coverage circle considerably larger than LEO and not much smaller than GEO. Its orbit requires more energy to attain than LEO but less than GEO. Required sensor range, while greater than for a LEO, is considerably less than for a GEO.

The MEO range is currently inhabited primarily by navigation satellites. The U.S. Global Positioning System (GPS or NAVSTAR) consists of 24 satellites in six orbit planes (four satellites per plane) inclined at 55 deg. The satellites are at an altitude of about 20,000 km, in a $Q = 2$ circular orbit, with a groundtrack that repeats each day. Figure 15.13 shows the groundtrack of a single GPS satellite and the coverage circle projected on the Earth at an instant in time. Note that the groundtrack repeats itself after two revolutions (12-h orbit period) and that the size of the coverage circle is not much smaller than for a GEO satellite, even though the altitude is nearly half that of GEO. The purpose of this satellite system is to provide at least fourfold continuous Earth coverage to allow a user to determine his position accurately. The Russian GLONASS (Global Navigation

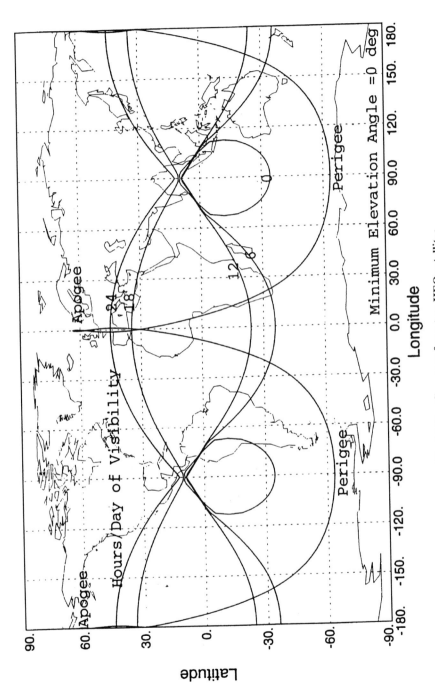

Fig. 15.11   Coverage from two HEO satellites.

coverage are examined for as many as 100 to 200 satellites. Later tables by these same authors investigate constellations of up to several thousand satellites. For as many as 100 satellites, the number of satellites required for continuous global coverage as a function of satellite altitude (h) for these optimally phased, non-symmetric, polar constellations of satellites using the street of coverage method is shown in Fig. 15.18. As expected, the number of satellites increases steadily (although not always monotonically) as the satellite altitude decreases. Note that twofold coverage of the globe does not require twice as many satellites as onefold. Sometimes, an additional fold of coverage is available for a minor percentage increase in the number of satellites.

Optimal Satellite Constellations Using Walker's Method

Researchers such as Walker,[10, 11, 14–16, 18] Mozhaev,[12, 13] Ballard,[17] and Lang[19, 20] have sought constellations of satellites in common-altitude, generally inclined, circular orbits that provide continuous global (single or multiple) coverage with a minimum of satellites. Only symmetric arrangements of satellites are considered.

Using Walker's notation, symmetric constellations of satellites can be described by the parameters $T/P/F$ and i, where

T = total number of satellites in the constellation
P = number of commonly inclined orbital planes
F = relative phasing parameter
i = common inclination for all satellites

In order to have a symmetric arrangement, the T/P satellites in a given orbital plane are equally spaced in central angle (phasing), and the P orbital planes are evenly spaced through 360 deg of right ascension of ascending node. The phasing parameter F relates the satellite positions in one orbital plane to those in an adjacent plane (i.e., interorbit phasing). The units of F are $360/T$ deg.

Table 15.1   Initial satellite locations for example 12/3/2 constellation

Satellite No.	Plane No.	RAAN, deg	Initial mean anomaly, deg
1	1	0	0
2	1	0	90
3	1	0	180
4	1	0	270
5	2	120	60
6	2	120	150
7	2	120	240
8	2	120	330
9	3	240	120
10	3	240	210
11	3	240	300
12	3	240	30

Fig. 15.19 Walker 12/3/2 constellation (three planes separated by $\Delta\Omega$ = 120 deg, four satellites per plane, alt = 1000 km, incl = 60 deg).

As an example, consider $T/P/F = 12/3/2$. The 12 satellites are located such that four satellites are evenly spaced in each of three orbit planes. The three orbit planes are 120 deg apart in right ascension of ascending node (RAAN). If there is a satellite at its ascending node (mean anomaly = 0) in orbit plane 1, then, in orbit plane 2 (adjacent plane to the east), a satellite will be at a mean anomaly of $0 + F \times 360/T = 60$ deg. For this example, the arrangement of all 12 satellites in terms of right ascension of ascending node and initial mean anomaly is given in Table 15.1. An illustration of this Walker 12/3/2 constellation with an inclination of $i = 60$ deg is shown in Fig. 15.19.

The parameters $T/P/F$ and the orbital inclination i are sufficient definition of a constellation to allow the determination of the Earth central angle radius of coverage θ required for any specified level of continuous global coverage. To accomplish this, Walker devised the circumcircle approach. Using this method, the satellite positions are computed at small time steps. At each time step, all combinations of three satellites are examined, and the radius of the spherical circle that contains them (the *circumcircle*) is determined. Figure 15.20 shows the circumcircle approach applied at a single time step. The largest circumcircle over all time intervals that does not include another satellite is equal to the size of the coverage footprint θ necessary to achieve continuous global coverage. For double or twofold coverage, the largest circumcircle that includes one satellite is sought, and so on for multiple folds of coverage.

To optimize the constellation, Walker numerically varied the orbital inclination so as to minimize θ. With this done, the optimal value of θ was obtained for a specific set of $T/P/F$. For a specified value of T total satellites, this procedure must be repeated for all values of the number of orbit planes P (i.e., all the factors of T) and for all values of the phasing parameter F (for each value of P, F can take a value of 1 through P) in order to determine the overall best set of $T/P/F$ and i.

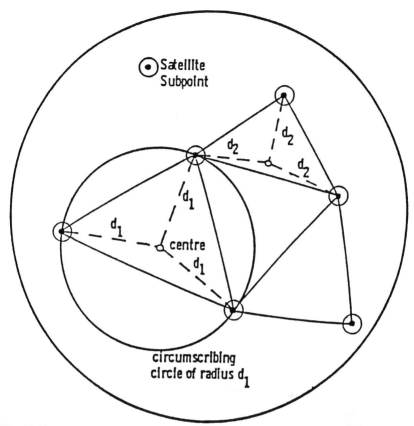

Fig. 15.20 Walker's circumcircle approach. At each instant in time the largest circumcircle defined by three satellite subpoints is determined.

Walker's circumcircle approach is numerically intensive because its accuracy depends on the size of the time step used. Additionally, as the number of satellites T increases, the number of combinations of three satellites used to define each circumcircle increases as $[T \times (T - 1) \times (T - 2)]$. For these reasons, this method is computationally demanding for large values of T. Walker was able to use this approach to optimize constellations of as many as 24 satellites for onefold through fourfold continuous global coverage. Lang[19] abandoned the circumcircle approach and used some symmetry simplifications to reduce the computational requirements for large constellations. In this manner, optimal Walker-type constellations have been determined for up to 100 satellites.

The resulting optimal Walker-type constellations are shown in Fig. 15.18 in terms of the number of satellites required to achieve the desired level of continuous global coverage as a function of satellite altitude. Optimal constellations for both the streets of coverage and Walker methods are included on this plot. Recall that the constellations from the streets of coverage technique involve optimally phased satellites in polar orbits spread over roughly 180 deg in node. Typically, the number of satellites per plane far exceeds the number of planes. The Walker technique

ORBITAL MECHANICS

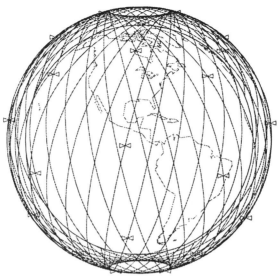

Fig. 15.21 Optimal Walker-type constellation for continuous global coverage ($T/P/F$ = 32/32/28) (32 planes separated by $\Delta\Omega$ = 11.25 deg, 1 satellite per plane, alt = 1048 km, incl = 76.9 deg).

produces symmetric constellations of inclined orbit planes with nodes distributed around 360 deg. Often, there is only one satellite per plane.

Figure 15.21 depicts the optimal 32-satellite Walker-type constellation for continuous global coverage. It employs 32 separate orbital planes inclined at 76.9 deg with one satellite per plane ($T/P/F$ = 32/32/28). The orbital altitude is 1048 km to achieve continuous global coverage with a minimum elevation angle of 0 deg. This altitude is nearly 150 km higher than the corresponding optimal polar constellation (see Fig. 15.17), thereby making the polar constellation the more efficient. For an orbit planner, however, efficiency may not be the only consideration in selecting between such constellations. Despite its higher altitude, the Walker-type constellation has a lower inclination (76.9 instead of 90 deg), which might make its orbit easier to achieve, depending on the launch vehicle and launch site. On the other hand, planners might prefer the polar constellation because they can minimize their launch costs by lofting two, four, or even eight satellites together to the same orbital plane using a single launch vehicle. In contrast, this particular Walker-type constellation would require a separate launch to each of the 32 orbital planes. Less optimal (higher-altitude) Walker-type constellations are likely to be found that use fewer planes to achieve the same coverage. Clearly, arriving at the most cost-effective constellation often involves a detailed tradeoff analysis.

In comparing the optimal streets of coverage and Walker-type constellations as shown in Fig. 15.18, several conclusions can be drawn. For single continuous global coverage with 20 or fewer satellites, the symmetric, inclined Walker-type constellations are more efficient. For the same number of satellites, they offer continuous global coverage at a lower altitude (correspondingly lower θ). Conversely, at the same altitude, they can perform the same job with fewer satellites.

For single continuous global coverage with more than 20 satellites, the optimally phased, nonsymmetric polar constellations (via the streets of coverage method) are more efficient. For double or higher folds of continuous global coverage, the Walker-type constellations are always more efficient. In fact, in the region of 30 satellites, the inclined Walker-type constellations achieve fourfold coverage at altitudes for which the polar constellations cannot even achieve full threefold coverage.

Optimal Satellite Constellations Using Draim's Method

Earlier, we saw how two or three highly eccentric orbit (HEO) satellites could be arranged in a single groundtrack with apogee at the northernmost point so as to provide concentrated coverage over the northern hemisphere. Similarly, placing apogee at the southernmost point in the orbit would concentrate the coverage in the southern hemisphere. Using elliptical orbit satellite constellations with apogees located half in the north and half in the south presents an alternate way of obtaining global coverage. Draim[21-24] has used this technique to arrange satellites in high-altitude, moderately elliptical, moderately inclined orbits in a polyhedral fashion so as to obtain single and multiple continuous global coverage. In this manner, he has achieved one-, two-, three-, and fourfold continuous global coverage with 4, 6, 8, and 10 elliptical orbit satellites, respectively. This is one satellite fewer than would be required by the best circular orbit constellations even at the highest altitudes. The orbital elements of the 4, 6, 8, and 10 satellite constellations are shown in Table 15.2.

The values given for semimajor axis are the lowest values that yield the specified fold of continuous global coverage using a ground elevation angle ε of 0 deg. Note that these Draim constellations are all supersynchronous; that is, the orbital periods (26.49 h for onefold, 102 h for twofold, 272 h for threefold, and 568 h for fourfold) are all greater than 24 h. Since the eight- and ten-satellite constellations operate at such high altitudes, the severe lunar perturbations limit the practical value of these constellations.

15.3   Considerations in Selecting Satellite Constellations

The foregoing discussions have investigated the coverage offered by satellites in different types of orbits and methods for arranging multiple satellites in constellations so as to optimize their collective coverage. Although the coverage provided from a satellite constellation is a strong driver in the constellation selection process, it may not be the primary driver. Other factors that influence constellation selection will be addressed in this section.

Most often, the ultimate goal of satellite constellation selection is to achieve the desired coverage task with a minimum overall system cost. Although this objective is simple to state, the process involved can be quite complicated. Figure 15.22 illustrates a typical constellation selection process. The selection process pictured here is simplified and highlights the role played by the orbital constellation in the overall mission. The mission objectives usually dictate the coverage required of the system (e.g., regional, zonal, global, continuous, part-time) and certain sensor constraints (e.g., minimum elevation angle, lighting conditions, maximum slant range). These requirements and constraints become the inputs to the constellation

Table 15.2   Satellite orbital elements for Draim global coverage constellations

Sat. no.	Semimajor axis, km	Eccentricity	Inclination, deg	RAAN, deg	Arg of periaosis, deg	Mean anomaly, deg
			Four-satellite, onefold continuous global coverage constellation:			
1	45,033	0.263	31.3	0	270	0
2	45,033	0.263	31.3	90	90	270
3	45,033	0.263	31.3	180	270	180
4	45,033	0.263	31.3	270	90	90
			Six-satellite, twofold continuous global coverage constellation:			
1	110,630	0.233	27.5	0	270	0
2	110,630	0.233	27.5	60	90	300
3	110,630	0.233	27.5	120	270	240
4	110,630	0.233	27.5	180	90	180
5	110,630	0.233	27.5	240	270	120
6	110,630	0.233	27.5	300	90	60
			Eight-satellite, threefold continuous global coverage constellation:			
1	212,742	0.218	25.0	0	270	0
2	212,742	0.218	25.0	45	90	315
3	212,742	0.218	25.0	90	270	270
4	212,742	0.218	25.0	135	90	225
5	212,742	0.218	25.0	180	270	180
6	212,742	0.218	25.0	225	90	135
7	212,742	0.218	25.0	270	270	90
8	212,742	0.218	25.0	315	90	45
			Ten-satellite, fourfold continuous global coverage constellation:			
1	347,567	0.205	24.0	0	270	0
2	347,567	0.205	24.0	36	90	324
3	347,567	0.205	24.0	72	270	288
4	347,567	0.205	24.0	108	90	252
5	347,567	0.205	24.0	144	270	216
6	347,567	0.205	24.0	180	90	180
7	347,567	0.205	24.0	216	270	144
8	347,567	0.205	24.0	252	90	108
9	347,567	0.205	24.0	288	270	72
10	347,567	0.205	24.0	324	90	36

selection process. Often, these inputs themselves may be changed as the process continues, perhaps as a result of improved technology, but most often because certain requirements or constraints prove too costly.

These inputs to the selection process are first used to establish a candidate satellite constellation. The constellation may be selected from the various methods described earlier (e.g., streets of coverage, Walker-type, Draim). By whatever method, the number of satellites in the constellation and their altitude, inclination, and number of orbital planes are determined.

15.2. a) ΔRAAN $= 360/P = 360/5 = 72$ deg
   b) ΔMean anomaly $= F \times 360/T = 1 \times 360/5 = 72$ deg

15.3. a) From Eq. (15.13),

$$\cos\theta = (\cos c)\cos\pi/s$$

for $\theta = 81.3$ deg, $s =$ three sats/plane, solving for the half-street width c gives

$$c = 72.4 \text{ deg}$$

b) Since $\theta < 90$ deg for any satellite viewing the Earth's surface, it takes at least three satellites to give a street of coverage.

15.4. a) From Eq. (15.1),

$$\sin\alpha = \frac{\cos\varepsilon}{1 + h/r_e}$$

for $\varepsilon = 0$ deg, $h = 1000$ km, $r_e = 6378$ km, we get

$$\alpha = 59.8 \text{ deg}$$

b) From Eq. (15.1),

$$\cos(\theta + \varepsilon) = \sin\alpha$$

$$\text{or} \quad \theta + \varepsilon + \alpha = 90 \text{ deg}$$

for $\varepsilon = 0$ deg, $\alpha = 59.8$ deg, we get

$$\theta = 30.2 \text{ deg}$$

c) From Eq. (15.4),

$$\rho^2 = r_e^2 + (r_e + h)^2 - 2r_e(r_e + h)\cos\theta$$

solving for ρ yields

$$\rho = 3708 \text{ km}$$

d) From Eq. (15.13),

$$\cos\theta = (\cos c)\cos\pi/s$$

for $\theta = 30.2$ deg, $s = 8$ sats/plane, solving for the half–street width c gives

$$c = 20.7 \text{ deg}$$

e) ΔRAAN $= 360/P = 360/33 = 10.9$ deg

15.5.
$$\frac{W_{\text{GEO}}}{W_{\text{LEO}}} = \frac{\rho_{\text{GEO}}^{0.5}}{\rho_{\text{LEO}}^{0.5}}$$

so that, for a LEO weight of 400 kg, the GEO weight will be

$$W_{\text{GEO}} = 1341 \text{ kg}$$

15.6.   The completed table follows.

	Satellite Constellation			
	LEO		GEO	
	Walker 33 sats 33 planes	Street of coverage, 32 sats, 8 planes	Walker 5 sats 5 planes	Street of coverage, 6 sats, 2 planes
Satellite costs				
Sat wt (kg)	400	400	1341	1341
Sat cost ($M/kg)	0.08	0.08	0.08	0.08
No. of sats	33	32	5	6
Total sat cost ($M)	**1056**	**1024**	**536**	**644**
Launch vehicle costs				
LVXS				
Payload capacity (kg)	450	400	—	—
LV cost ($M/launch)	10	10	—	—
No. of launches req'd	33	32	—	—
Total LV cost ($M)	**330**	320	—	—
LVMD				
Payload capacity (kg)	4000	3500	—	—
LV cost ($M/launch)	60	60	—	—
No. of launches req'd	33	4	—	—
Total LV cost ($M)	1980	**240**	—	—
LVLG				
Payload capacity (kg)	—	—	1450	1350
LV cost ($M/launch)	—	—	80	80
No. of launches req'd	—	—	5	6
Total LV cost ($M)	—	—	**400**	**480**
LVXL				
Payload capacity (kg)	—	—	4500	4300
LV cost ($M/launch)	—	—	280	280
No. of launches req'd	—	—	5	2
Total LV cost ($M)	—	—	1400	560
Total system cost *sat + launch vehicle*	**1386**	**1264**	**936**	**1124**

The most cost-effective Comsat approach in this example is the five-satellite Walker constellation, launched singly on the large launch vehicle (LVLG). Note that the solutions are fairly close and that changing some of the launch vehicle capabilities or costs could significantly affect the answer.

Index